STATISTICAL THINKING IN SPORTS

STATISTICAL THINKING IN SPORTS

EDITED BY
Jim Albert and Ruud H. Koning

CRC Press
Taylor & Francis Group
Boca Raton London New York

CRC Press is an imprint of the
Taylor & Francis Group, an **informa** business

CRC Press
Taylor & Francis Group
6000 Broken Sound Parkway NW, Suite 300
Boca Raton, FL 33487-2742

First issued in paperback 2019

ISBN-13: 978-1-58488-868-0 (hbk)
ISBN-13: 978-0-367-38885-0 (pbk)

Library of Congress Cataloging-in-Publication Data

Statistical thinking in sports / editors, Ruud H. Koning and James H. Albert.
    p. cm.
  Includes bibliographical references and index.
  ISBN-13: 978-1-58488-868-0 (alk. paper)
   1. Sports--Statistical methods. 2. Sports--Statistics. I. Koning, Ruud H. II. Albert, Jim, 1953-

GV741.S825 2007
796.02'1--dc22
                                                               2007014052

Preface

The inspiration for this book begins with sports and its associated statistics. All sports contests are played with a particular numerical scoring system; one counts goals in soccer, points in basketball, and runs in baseball and cricket. Great accomplishments by professional athletes are measured by their statistics. These statistics are used to compare and rank athletes and to set salaries. The public has a fascination for the statistics in sports. In fact, these statistics have taken a life of their own with the recent introduction and growth of fantasy sports, where people draft players and games are won or lost solely on the basis of "real-life" statistics.

Research in statistical thinking in sports has a long history. Early collections of papers in statistics and sports appeared in books such as *Management Science in Sports* and *Operations Strategies in Sports* in the 1970s. The more recent volume *Statistics in Sport*, published in 1998, provides a broad survey of statistics research in sports, including individual chapters on American football, baseball, basketball, cricket, soccer, golf, ice hockey, tennis and track and field. A number of journals, such as *Chance*, *The American Statistician* and *The Statistician*, have regularly featured articles on the statistical analysis of sports data. The volume *Anthology of Statistics in Sports* contains a collection of the "best" sports articles that have appeared in journals sponsored by the American Statistical Association. Recently, the *Journal of Quantitative Analysis in Sports* was launched in 2005 to be the first academic journal devoted to the statistical analysis of sports. There are now more publication opportunities for statisticians with a serious interest in questions in sports.

A number of research groups have been formed that are dedicated to sports statistics. The American Statistical Association began a Section on Statistics in Sports in 1992. This group sponsors invited and contributed paper sessions at the annual Joint Statistical Meetings and maintains a web site. Also in 1993 the International Statistical Institute created a Sports Statistics Committee that also publishes proceedings at their bienniel meetings.

The intent of this volume is to provide an accessible survey of current research in statistics and sports. The chapters cover a range of sports issues of interest to an international audience written by experts in the field. All of the authors have attempted to write their chapters to be understandable by readers with a passing knowledge of the particular sport. The primary purpose of the articles is to explain how statistical thinking can be used to answer interesting questions on a particular sport. Although the articles are aimed at general readers, the serious researcher in sports statistics may also find the articles interesting and use them as starting points for further research.

We thank the following persons for their help. Chapters in this volume have been commented upon by Jos de Koning, Bart Los, Alec Stephenson, Tom Wansbeek, Ryanne van Dalen, Jay Bennett, Eric Bradlow, and Chris Andrews. Technical support has been provided by Siep Kroonenberg and Sashi Kumar. Finally, it was very pleasant to work with the editorial staff of Taylor and Francis.

Jim Albert   Ruud H. Koning
Bowling Green State University   University of Groningen
USA   The Netherlands

Contents

1

Introduction

Jim Albert

Bowling Green State University

Ruud H. Koning

University of Groningen

1.1 INTRODUCTION

Sports have taken an ever more prominent position in society. An increasing number of people watch sports events on television and more people see live sports in stadiums and arenas. The economic value of franchises, broadcasting rights, and merchandising has grown. Books on sports appear on the *New York Times* bestseller list (for example, Lewis, 2004). Besides this increasing interest from the general public, scientists have also taken on sports for their research agenda. Traditionally, research from physiology and medicine has been used to improve performance in sports. Currently researchers from economics, statistics, sociology, and law are working in the sports field.

In this volume, the focus is on the statistical analysis of sports. There has always been a close connection between sports and statistics. In most sports, players and teams are measured by various statistics, and these statistics are used to provide rankings of players and teams. A recent popular phenonema is fantasy sports where participants draft teams of players and games are won and lost on the basis of actual statistical information.

One reason for the close connection between statistics and sports is probably the abundant data that is available on sports. Scores are kept and individual performance is measured and tracked over time. The advent of the internet has perhaps helped to distribute these data to an ever wider group of researchers. Also, sport is a convenient and familiar context to use in teaching or in demonstrating a new statistical method.

In this volume, we illustrate a number of different models for sports data, such as time series, linear regression, ordered probit regression, factor analysis, and generalized linear models. A variety of distributions are used to model the variable of interest such as binomial, Gamma, Poisson, and others. Parameters are estimated by least squares, maximum likelihood, and Bayesian methods and used in simulation models. Indeed, sports provide a very broad area of application of statistical thinking.

1.1.1   PATTERNS OF WORLD RECORDS IN SPORTS (TWO CHAPTERS)

One fascinating subject for study is the pattern of world records in sports over time. Gerald Kuper and Elmer Sterken, in "Modelling the development of world records in running", provide an interesting look into the pattern of world records of metric running events for men and women. As the authors explain, there are interesting questions associated with this data. Can one estimate the ultimate human performance in these events? What is the impact of technology innovations on the pattern of records? Is the pattern of world records similar for different running lengths, and will women outperform men in the future? The authors provide a comprehensive survey of the use of different parametric families to model world records and give some interesting conclusions.

Ray Stefani in "The physics and evolution of Olympic winning performances" takes a "holistic" view of the pattern of winning performances over time in a variety of Olympic sports. To understand the changes in the winning performance in a given sport, say swimming, one should understand the factors influencing the power of a swimmer, such as the size and fitness of the athletes, the arm and leg positioning techniques, coaching, and the quality of the venue and equipment. Although one cannot directly compare the time of a swimming event with the height of a pole vault, Stefani defines a dimensionless measure, percent improvement per Olympiad, to describe the change in winning performances. This measure is used to contrast the evolution of winning performances in a wide variety of sports.

1.1.2   COMPETITION, RANKINGS, AND BETTING IN SOCCER (THREE CHAPTERS)

Soccer is the most popular sport in the world, as judged by the number of people playing or watching the sport. Besides being interesting as a sport, it has also become an economic activity of some significance. For example, anti-trust regulators watch the sale of television rights, and the European Commission is involved in setting up a system of transfer fees. Also, just as in many other sports, betting on soccer matches has become increasingly popular. Despite ongoing commercialization of soccer, one would almost forget that it is a game; and organizing leagues want to know which team is best. These issues are addressed in Chapters 4, 5, and 6.

Marco Haan, Ruud Koning, and Arjen Van Witteloostuijn look at the development of competitive balance over time in their chapter "Competitive balance in national European soccer competitions." They do so for seven different countries. First, they discuss different dimensions of competitive balance and propose empirical measures that capture these dimensions. Then they proceed to examine whether balance has changed over time, in particular, they investigate the popular belief that competitive balance has worsened over time. Finally, noting the lack of agreement on a single measure of competitive balance in soccer, they use a factor model to see whether seven different indicators can be reduced to one factor. It turns out that the predominant factor can be interpreted as contemporaneous competitive balance.

Soccer is played at different levels: club teams play in national leagues and international tournaments as the Champions League, and national teams play every four years to win the World Cup. Ranking of club teams is relatively easy, considering the number of games they play. It is much harder to rank national teams as, in a given year, they play only a limited number of games, against a selected set of opponents. Still, the world soccer federation FIFA publishes a ranking of national teams, that is updated frequently. What is the quality of this ranking? This issue is addressed by Ian McHale and Stephen Davies in "Statistical analysis of the effectiveness of the FIFA World Rankings." They conclude that the FIFA World Ranking does not use all past information efficiently.

Betting on sport results is a hobby for one, and a way of earning a living for others. To what extent are betting markets efficient? Stephen Dobson and John Goddard's "Forecasting scores and results and testing the efficiency of the fixed-odds betting market in Scottish league football" examine different betting strategies using two different statistical forecast models: a goals-based model and a results-based model. These forecasting models are capable of eliminating almost all of the bookmakers over-round.

1.1.3 AN INVESTIGATION INTO SOME POPULAR BASEBALL MYTHS (THREE CHAPTERS)

Baseball has been called the most statistical sport in the sense that more numerical information is collected about this game than any other. For a given baseball play, such as a batted ball hit into center field for a single, many associated variables will be recorded about the event including the inning, runners on base, the players on the field, and the exact location of the hit in the field. Websites such as The Baseball Archive (www.baseball1.com) and Retrosheet (www.retrosheet.org) provide extensive datasets on historical players and teams and play-by-play game results. The easy availability of this data invites interesting analyses by researchers that are reflected in the three baseball papers in this volume.

All three articles investigate the validity of popular myths in baseball. "Hitting in the pinch" by Jim Albert investigates the popular belief that particular ballplayers have the ability to perform better in important or "clutch" situations during a game. In his paper, Albert shows that the ability to hit well can depend on the runners on base and the number of outs in an inning. But there is little evidence to suggest that particular players have the ability to do better in important situations. Another popular belief in sports is the importance of momentum during a game. If particular players perform well during a game, many people believe that this will motivate other players to also perform well, causing the team to rally. Rebecca Sela and Jeffrey Simonoff take a statistical view of this issue in "Does momentum exist in a baseball game?" They begin with a Markov Chain model for baseball, where the probability of a movement from one state (say no runners on and one out) to another state (runner on first and one out) only depends on the beginning state. They consider a more sophisticated model where the probability of a movement can depend on various "momentum" variables. The authors find little statistical evidence of momentum effects, especially from a predictive viewpoint. Hal Stern and Adam Sugano in "Inference about batter-pitcher matchups in baseball from small samples" investigate the final myth, the importance of batter/pitcher matchup data. Baseball managers will often make decisions on the basis of how particular batters perform against particular pitchers. The problem is that one has many samples of large number of batter/pitcher matchups, and one is likely to see extreme sample outcomes by chance. Stern and Sugano suggest modeling this type of data by a hierarchical model; this model will allow one to see how a batter's ability can vary depending on the quality of the pitcher.

1.1.4   UNCERTAINTY OF ATTENDANCE AT SPORTS EVENTS (TWO CHAPTERS)

Professional sports is a business and all teams wish to have high attendance at their games. A popular assumption is that audiences will be attracted to games where the outcome is very uncertain. Babatude Buraimo, David Forrest, and Robert Simmons, in "Outcome uncertainty measures: how closely do they predict a close game?" note that there is little empirical support for this assumption. They suggest that one problem is that although there are several measures of outcome uncertainty used in the literature, it is unclear whether any of these measures are actually good predictors of close contests. This paper defines several measures of outcome uncertainty and finds that they only explain a small amount of the variation in game results of Spanish football.

Attendance at sports events is also a central theme in the article "The impact of post-season play-off systems on the attendance at regular season games" by Chris Bojke. Many different sports have introduced play-off systems; one benefit of these systems is that they enhance attendance by increasing the time that teams are in contention for the league championship. Unfortunately, there is little research on the impact of the play-off design on attendance and Bojke presents a statistical model to understand this relationship. By fitting the model to English football data, one is able to measure the impact of a particular play-off system on attendance. Even though his application is to soccer, the methodology can be applied to modeling attendance for other sports as well.

1.1.5   HOME ADVANTAGE, MYTHS IN TENNIS, DRAFTING IN HOCKEY POOLS, AMERICAN FOOTBALL

One is familiar with the saying "there is no place like home", and this statement is especially true for sports competitions. For all sports, the team playing in its home field or area generally has an advantage. Ray Stefani's article "Measurement and interpretation of home advantage" explores the home-field advantage for a number of sports. A mathematical model is used to quantify home field advantage for a particular sport and the size of the estimated home advantage is shown to differ greatly between sports. Stefani describes the physiological, psychological and tactical factors implicit in home-field advantage and argues that player fatigue, especially in a continuous-action sport, plays an important role in home advantage.

Jan Magnus and Franc Klaassen provide a nice survey of their tennis research in "Myths in tennis." Tennis is an international sport watched by fans all over the world and the television commentators have strong views about competition. In particular, commentators believe a player has an advantage if he or she serves first, and top players have a special ability to perform well in the important points in a match. This article explores these beliefs by use of four years of point-by-point tennis data from the Wimbledon tennis championships. Many interesting conclusions are reached about tennis matches and most of the beliefs held by television competitors are shown to be false.

David Berri's article "Back to back evaluations on the gridiron" describes the evaluation of player performance in American football. Baseball is a relatively easy sport in evaluating player performance since the game is essentially a confrontation between a single pitcher and a single batter. Football is fundamentally different from baseball in that the performance of a particular player such as a quarterback is highly dependent on the performance of his teammates on the field. Berri applies the regression methodology of Scully and Blass to develop a measure of marginal performance for football players. This methodology leads to some interesting measures of player performance; in particular, Berri's measure of a quarterback's performance is likely superior to the official quarterback rating system of the National Football League. But these measures do not appear to show consistency over time, suggesting that the statistics collected by professional football are not useful for measuring the productivity of individual players.

Amy Summers, Tim Swartz, and Richard Lockhart in "Optimal drafting in hockey pools" considers how one can be successful in a hockey fantasy league. In this game, participants draft players from the 16 National Hockey League teams that have qualified for the Stanley Cup Playoffs. The winner of the league is the one whose players have accumulated the largest number of points. An interesting statistical model is devised for the number of points scored by hockey players, and this is used to find an optimal selection strategy. One attractive by-product of this model is that a player can make intelligent draft choices in real time.

1.2 WEBSITE

This book is intended for sports enthusiats, with some background in statistics. They can be students, teachers, researchers, but also practitioners or (sport) policy makers. To promote more statistical thinking in sports, we have made a website `www.statistical-thinking-in-sports.com` that has additional material as appendices, references, tables, and other data. Feel free to use the information provided there, but do send a copy of your paper or project for inclusion on the website if you use material from that website.

REFERENCES

Lewis, M. (2004). *Moneyball: The Art of Winning an Unfair Game*. New York: W.W. Norton & Company.

2

Modelling the development of world records in running

Gerard H. Kuper
University of Groningen

Elmer Sterken
University of Groningen

ABSTRACT

We model the development of world records of metric running events from the 100 meter dash to the marathon for men and women. First, we review methods to fit time-series curves of world records in general. We discuss methods to estimate curves and review candidate functional forms that fit the systematic shape of the progress of world records. Next, we fit the asymmetric Gompertz curve for 16 events and compute implied limit values. In order to assess the implied limits we use the Francis (1943) model to relate limit records and distance in a log-log specification. We compare men and women and conclude that there is a fixed difference in record times between the two sexes. Finally, using the log-log relationship between time and distance we calculate the development of the world record of the mile as a robustness check.

2.1 INTRODUCTION

Over the past 50 years multiple studies have made attempts to model the development of world or Olympic records, for instance for running events. Deakin (1967) explored progress in the mile record, Chatterjee and Chatterjee (1982) for the 100, 200, 400, and 800 meter in the Olympic Games, Blest (1996) for running distances of 100 meter to the marathon, and recently Nevill and Whyte (2005) for 800 meter to the marathon by male and female runners. Modelling world records through time attracts attention from different perspectives. First, all studies are inspired by the apparent problems in the analysis of world records. In terms of time used to complete an event for instance we only observe nonpositive changes. These nonpositive changes can be very infrequent: world records sometimes survive for about 20 to 25 years. But in some instances improvements are really substantial, leading to extreme values in the distribution of the first difference of the series. Or technological innovation of sports gear shifts up the human frontier; an example is the clap skate in speed skating (see Kuper and Sterken, 2003 and Stefani, Chapter 3 of this volume). A second element in the analysis is the interest in the ultimate human performance. Given the development of the world record up to now, can we predict the fastest time ever? And thirdly, can we compare contemporaneous performances? How does the world record 10000 meter running for men compare to the 5000 meter for men? Is there a phase difference in the development of records of the various events? And can women outperform men in the far future?

Observing the time series of the world record of a well-developed sports event reveals an inverted S-shape pattern (we will treat all developments of records as monotonic declining time series of the time used to complete an event). In the early phase of the development of running events, competition is not fierce, and amateurism dominates. At the inflection point the rate of progress is large, because more sportsmen get involved, more professional help is available, rewards become more visible, etc. After this rapid development phase there is a phase of saturation. It is hard to improve the record and only at a few instances a highly talented individual is able to break it. For some sports events we do not observe such a shape of the time series, because the development is much faster through cross-fertilization (e.g., the 3000 meter steeple for women, which has become an official race distance only recently). For some events typical observations are available for some pieces of the curve. This sometimes leads to the use of simple piecewise linear techniques to model the development of world records. These linear approximations are computationally attractive, but theoretically poor. Nevill and Whyte (2005) e.g., contribute to the debate on whether linear approximations are helpful in describing world records. Tatem, Guerra, Atkinson, and Hay (2004) use linear approximations of the development of the best male and female 100 meter sprints at the Olympic Games and conclude that in 2156 women will run as fast as men. Whipp and Ward (1992) also employed a linear approximation of marathon times of men and women and predicted in 1992 that female marathon runners would run as fast as men in 1998. Of course linear approximations cannot be correct: world record times cannot become negative. Wainer, Njue, and Palmer (2000) argue that women show similar development patterns of world records, but lag the male equivalents. Their conclusion is that the growth rates of development of record times is equal between men and women. Our main question though is to find more biologically sound and statistically robust nonlinear (S-shaped) functions that provide a superior fit of the development of world records (see Nevill and Whyte, 2005).

Besides finding the best fit, most studies are trying to get insight into the upper-limit of speed (of running) or the lower limit of time to be used to complete an event. For some functional forms, a nice limit value can be derived from the properties of the functions. For instance Kuper and Sterken (2003) give detailed lower limits of time to be used on skating events, while Nevill and Whyte (2005) give predicted peak world records for the men's 800 meter, 1500 meter, mile, 5000 meter, 10000 meter, marathon and women's 800 meter, and 1500 meter. For skating the predictions of future world records are highly dependent on the "no-technical progress" assumption. Contrary to running, speed skating is a technology-intensive sport (skates, ice rinks, clothing), so that shocks to technological progress are visible in the improvements of world records. Ultimate human performances are so conditional on this typically hard to predict factor.

In this paper we contribute to the literature on modelling world records. We focus on running events, since these events are well-developed, highly competitive, not intensively affected by the problem of hard to predict technical innovations, and have a typical long history. For instance for the one mile record we have official data since 1865 (for men). Firstly, we compute the approximation of historical lower bounds of time needed to complete running events using the single-event historical data. After a careful selection and discussion of alternative specifications, we apply the Gompertz curve (Gompertz, 1825). The Gompertz curve is a relatively simple curve that allows for an asymmetric S-shape. From these specifications we compute the implied infinite lower bounds. Secondly, these lower bounds are compared in a cross-sectional setting to find the relationship between time and distance. Finally, we use one event, the one mile run, to compare the forecasting performance of our methodology.

The set-up of the chapter is as follows. First, we discuss the existing methodology to model world records in Section 2.2. In Section 2.3 we present different functional forms. From this analysis we conclude to use the Gompertz model. Section 2.4 describes the development of world records in running. In Section 2.5 we present the results of fitting the Gompertz curve for the 100, 200, 400, 800, 1500, 5000, 10000 meter, and marathon events for men and women. In Section 2.6 we test for robustness of the methods by relating the limit values implied by the Gompertz curves and distance in the famous log-log model of distance and time (see Section 2.2 for a description). We summarize and conclude in Section 2.7.

2.2   MODELLING WORLD RECORDS

In this section we present two approaches to model world records. First we review the literature on comparing world records at the same moment of time on various distances. Next we more extensively discuss candidate functional forms to fit the historical development of world records. Before doing so, we first discuss in short the problems in estimation. How do we cope with the extreme nature of world records? For instance Kuper and Sterken (2003) estimate a so-called Chapman-Richards approach (see hereafter for a precise definition) to the development of the trend using normally distributed residuals and calculate the extreme residual to compute the shift to the extreme frontier. Smith (1988) proposes a maximum likelihood method of fitting a model to a series of record times. Let Y_t be the best performance in a particular event in the t-th year:

$$Y_t = X_t + y_t,$$

where X_t is an iid-variable and y_t a nonrandom trend. The records in a T-year period are given by:

$$Z_t = \min(Y_1, \ldots, Y_t),$$

for $1 \leq t \leq T$. The sequence Z_t is observable, but the underlying data Y_t is not. We concentrate on two parameterizations. First, we can think of the density function

$f(x; \theta)$ of X_t. Let $F(x; \theta)$ be the cumulative density function of $f(x; \theta)$. The statistical problem is then a censored data problem in which the value of Y_t, in a nonrecord year, is censored at the record level. Smith (1988) analyses the normal distribution (which we do not describe here), the so-called Gumbel extreme-value distribution:

$$F(x; \mu, \sigma) = 1 - \exp[-\exp((x - \mu)/\sigma)],$$

and the generalized extreme-value (GEV) distribution:

$$F(x; \mu, \sigma, k) = 1 - \exp\left[-(1 + k(x - \mu)/\sigma)^{\frac{1}{k}}\right],$$

The GEV with $k \to 0$ gives the Gumbel distribution. For $k \geq 1$ the GEV-function does not yield a local maximum. Smith indeed finds that these extreme-value functions have fitting problems and proceeds with the normal distribution. Sterken (2005) applies a similar approach to age-dependent running performance data in a stochastic frontier analysis. Stochastic frontiers allow for measurement error in the data, but assume that a large fraction of the deviation of observed records to the frontier is due to inefficiency. The stochastic frontier analysis can also be seen as an alternative approach to trim extreme values.

The main attention in this paper focuses on the modelling of the trend y_t though. We review the ideas on the parametrization of y_t hereafter. But first, we review the relationship between y_t-limit values for various distances.

2.2.1   CROSS-SECTIONAL APPROACH

We can analyze the progress of world records for different events separately, but we could also consider cross-sectional evidence. For instance, for running Francis (1943), Lietzke (1954), and Grubb (1998) analyzed the log-log relationship between running time and distance. Kennelly (1905) stated the relationship between time t and distance d after a study of race horses and human athletes as

$$\log t = 9/8 \log d - \text{constant}.$$

Define velocity $v = d/t$. Francis (1943) considered the relation:

$$(\log d - 1.5)(v - 3.2) = 6.081,$$

and found that the 5000m record in 1943 was a sign of high performance. In this model the $\exp(1.5)$ is an asymptote, which gives the distance at which the maximum speed is attained. Rewriting the Francis equation in general terms $v = A/[\log d - B] + C$, Mosteller and Tukey (1977) proposed the relation:

$$v = A(d - B)^\lambda + C,$$

where C is the speed at long distances, B the distance at which the maximum speed is obtained, and A the decrease in speed with transformed distance.

Lietzke (1954) noted that, starting from the log-log relation between time used and distance:

$$\log d = k \log t + \log a, \tag{2.1}$$

we get $t = (d/a)^{1/k}$. This implies that $v = d/t = a^{1/k} d^{(k-1)/k}$. So:

$$\log v = \frac{k-1}{k} \log d + \frac{1}{k} \log a.$$

Lietzke labels the constant $(k-1)/k$ as the exhaustion constant and estimates different values for running, horse racing, swimming, cycling, walking, and even auto racing.

Blest (1996) estimates a model for the Olympic records t_{ij} for the distance d_j at the time of the i-th Olympiad:

$$t_{ij} = A_i d_j^{\gamma_i},$$

where A_i and γ_i are parameters to be estimated. An alternative model is $t_{ij} = A d_j^{\gamma_i}$, where A is an average value over n Olympiads:

$$A = \exp\left(\frac{1}{n} \sum_{i=1}^{n} \log A_i \right).$$

No matter what model is used, there seems to be a clear log-log relation between speed and distance, or running time and distance. We explore this relation for the limit values of the Gompertz curves of individual distances that we compute hereafter. In order to find the limit values we need to describe the curves of individual world record time series.

2.2.2 FITTING THE INDIVIDUAL CURVES

In this subsection we briefly review the methodology of finding functional forms of the nonrandom trend y_t that are or could be used in fitting the progress of world records. In Section 2.3 we review in detail the various candidate functions and present our selection. Most of the work on selection of functional forms is done in the fields of medicine and biology. For instance, forest growth and growth rates of population, organs, and bones are described by the candidate functional forms discussed here. The main difference between the biology/medicine models and our data is the discrete-jump nature of the world record time series. Moreover, in biological growth models one can observe growth from birth until death, while our time series only reveal a small piece of the whole statistical process. Despite these differences the similarities are remarkable.

The most common way to express growth models is by describing the differential dy/dt, the rate change of y with respect to time t. For those variables that are expected to have a limit value at time T, stationarity after t implies $dy/dt = 0$. Up to time t dy/dt can take different functional forms. Some of these functional forms depend on growth theories. The most famous example is the Malthusian growth model $dy/dt = ry$ (see Malthus, 1798, and the exponential growth model hereafter). Some of the growth models assume a sudden stop of growth at the final date. These models can vary in the growth rate functional form until T: is the growth rate increasing, constant, or decreasing? Another approach is to assume that before T dy/dt is decreasing, implying a hump-shaped growth form. The growth rate increases, reaches a maximum, and decreases again. Integrating dy/dt differentials leads to the time path, and the other way of representing growth of y_t. We show examples below. Some of the growth models are based on growth assumptions, some other models find their origin in curve fitting exercises. We will mix both approaches, but denote the growth-theoretical assumptions when present.

2.3   SELECTION OF THE FUNCTIONAL FORM

In this section we first present the different functional forms and then discuss our procedure to select a standard curve to be used in fitting running world records. One could use different curves for different events, but we want to compare the limit values hereafter and so use a single curve. Below we will use the following four main parameters to describe the growth functions. We denote the units of observation y_t as the time needed to complete the distance. θ_1 will be used as the symbol of the limit value of y_t if $t \to \infty$. We use θ_2 as the indicator of growth. θ_3 will be used as the starting value of y_t at the starting date. θ_4 will be used as the indicator of time at the inflection point (see hereafter). Other parameters will be denoted by higher indices.

2.3.1   CANDIDATE FUNCTIONS

2.3.1.1   LINEAR TREND

Assuming a constant growth rate of performance, denoted by the simple differential equation $dy/dt = -\theta_2$, and taking the integral, the linear trend $y_t = \theta_3 - \theta_2 t$ is one of the first candidates to use (see Whipp and Ward, 1992, and Tatem et al., 2004). Although a linear trend cannot be a valid model of lower bounds in the long run (there is no limit value of e.g., speed), local linear approximations often perform rather well in curve fitting. So if the focus of the fitting is purely descriptive (and not on forecasting or computing limit values) local linear approximations, e.g., data envelopment analysis (see Charnes, Cooper, and Rhodes, 1978), are true candidates.

2.3.1.2   EXPONENTIAL CURVE

The famous Malthusian population growth model is an exponential model. In order to get a limit we can use the standard growth model, that can be described by the following linear autonomous first-order differential equation:

$$\frac{dy}{dt} = \theta_2(\theta_1 - y).$$

Solving this simple differential equation yields:

$$y_t = \theta_1 + (\theta_3 - \theta_1)\exp(-\theta_2 t).$$

The exponential function $y_t = \theta_1 + (\theta_3 - \theta_1) \exp(-\theta_2 t)$ is the mostly used asymptotic model (see e.g., Deakin, 1967). For positive values of θ_2 the limit value of the series is θ_1. Ratkowsky (1990) describes this function as a close to linear model, since the estimates of the parameters encompass the ones of a linear model. The main critique on the model is its monotonic change over time, which is mostly not apparent in standard S-shaped (sigmoid) curves. So there is no point of inflection, which is mostly found in graphs of the development of world records.

The exponential function is sometimes represented as the specialized Von Bertalanffy function (von Bertalanffy, 1938), which is widely used to describe e.g., the growth of a variety of animals:

$$y_t = \theta_1 + (\theta_3 - \theta_1) \exp(-\theta_2(t - \theta_5)),$$

where θ_5 is the more general notation of the starting time. The Von Bertalanffy growth model is hard to fit. In biology growth is often plotted in a Ford-Walford plot. This plot graphs the equation in a phase diagram:

$$y_{t+1} = -\theta_1 \exp(-\theta_2) + y_t \exp(-\theta_2).$$

Regressing y_{t+1} on y_t gives estimates of $\exp(-\theta_2)$ and so of θ_1 as well. A problem with the time series of world records is that most of the $y_{t+1} - y_t$-observations are equal to zero, which complicates estimation of a Ford-Walford plot.

2.3.1.3 MODIFIED WEIBULL CURVE

The modified Weibull $y_t = \theta_1 + (\theta_3 - \theta_1) \exp(-\theta_2 t^{\theta_5})$ function is also in the exponential class and has limit value θ_1 for a growth parameter $\theta_2 > 0$. The Weibull function is a very flexible function and widely used to model growth and yield data. The parameters θ_2 and θ_5 are scale and shape parameters. For $\theta_5 = 1$ we get the basic exponential form.

2.3.1.4 ANTISYMMETRIC EXPONENTIAL FUNCTION

Most of the functions have a symmetric S-shaped function for certain parameter combinations. Blest (1996) and Grubb (1998) propose the antisymmetric exponential function:

$$y_t = \theta_1 + (\theta_3 - \theta_1) \exp[-\theta_2(t - \theta_4)] \text{ if } t \geq \theta_4,$$
$$y_t = \theta_1 + (\theta_3 - \theta_1)[2 - \exp(\theta_2(t - \theta_4))] \text{ if } t < \theta_4.$$

This function has the positive limit θ_1 and allows for asymmetry.

2.3.1.5 HILL CURVE

Hill (1913) proposed the following curve:

$$y_t = \frac{\theta_3 \theta_2^{\theta_5} + \theta_1^{\theta_5}}{\theta_2^{\theta_5} + t^{\theta_5}}.$$

This curve is not based on a theoretical growth model and does not have a sensible limit value. Note that θ_3 is the value of y_t at the start of the series and θ_1 the value of y_t at the end of the series, and so not the limit values. The Hill-curve is more a descriptive curve, than a curve that can be used to make predictions of the development of world records.

2.3.1.6 CHAPMAN-RICHARDS' CURVE

The exponential class of growth functions is a prominent one. Above we showed that the simple Von Bertalanffy model can be considered in this class. Next we discuss a generalization of the model. It is good to denote how a generalized Von Bertalanffy equation originates from a growth model. As explained for the exponential form the standard growth model is:

$$\frac{dy}{dt} = \theta_2(\theta_1 - y),$$

where the growth rate θ_2 and limit value θ_1 are constants. More flexibility is obtained by substituting a power transformation y^{θ_5} for y (with θ_5 unequal to 0 or 1):

$$\frac{dy^{\theta_5}}{dt} = \theta_2\left(\theta_1^{\theta_5} - y^{\theta_5}\right).$$

So

$$\theta_5 y^{\theta_5-1}\frac{dy}{dt} = \theta_2\left(\theta_1^{\theta_5} - y^{\theta_5}\right),$$

or:

$$\frac{dy}{dt} = \frac{\theta_2}{\theta_5}y\left[\left(\frac{\theta_1}{y}\right)^{\theta_5} - 1\right].$$

Define $\eta = \theta_2\theta_1^{\theta_5}/\theta_5$, the allometric constant $m = 1 - \theta_5$, $r = \theta_2/\theta_5$, and we get:

$$\frac{dy}{dt} = \frac{\theta_2\theta_1^{\theta_5}}{\theta_5}y^{1-\theta_5} - \frac{\theta_2}{\theta_5}y.$$

This equation, sometimes labeled as the Richards growth model, says that the growth at time t is assumed to be proportional to the size at time t multiplied by a saturation function (see Richards, 1959). Under specific assumptions of parameter values this equation can be integrated to:

$$y_t = \theta_1\left[1 - \left(\frac{\theta_1^{\theta_5} - \theta_3^{\theta_5}}{\theta_1^{\theta_5}}\right)\exp(-\theta_2 t)\right]^{\frac{1}{\theta_5}},$$

where θ_3 is the starting value of y_t at $t = 0$. For $\theta_5 \to 0$ this model approaches the Gompertz curve (see hereafter).

We will simplify things and jump to a specification of the Richards model that is called the Chapman-Richards model. This model can allow for a so-called sigmoid form and have a point of inflection (but does not need to have it in all parameter settings, see hereafter). With the logistic and Gompertz model (see both hereafter) the Chapman-Richards model is a popular model of growth dynamics. The Chapman-Richards model reads:

$$y_t = \theta_3 + (\theta_1 - \theta_3)[1 - \exp(-\theta_2 t)]^{\theta_5},$$

and contains a number of special cases. If $\theta_5 = 1$ we get the so-called natural growth (exponential) model. If $\theta_5 = -1$ we get a logistic model (see also hereafter):

$$y_t = \theta_3 + \frac{\theta_1 - \theta_3}{1 - \exp(-\theta_2 t)}.$$

The Chapman-Richards function has been used by Grubb (1998) for running and Kuper and Sterken (2003) for skating data. The limit value in the above specification is θ_3.

2.3.1.7 GOMPERTZ CURVE

Suppose that the rate of change can be described by:

$$\frac{d \log(y)}{dt} = \theta_2 \left(\log(\theta_1) - \log(y)\right).$$

Integration then leads to:

$$y_t = \theta_1 \exp\left(\exp\left(-\theta_2(t - \theta_4)\right)\right),$$

where θ_1 is again the upper limit, θ_2 the growth rate, and θ_4 the time at the point of inflection. This is the foundation of the Gompertz curve. The Gompertz curve is a special case of the logistic curve (see hereafter). The curve is developed by Gompertz for the calculation of mortality rates. Gompertz (1825) proposed the following curve:

$$y_t = \theta_1 + \theta_3 \exp\left(-\exp(\theta_2(t - \theta_4))\right).$$

Similar to the logistic function the Gompertz curve has asymptotes at θ_1 and $\theta_1 + \theta_3$. The inflection point is at $t = \theta_4$. The Gompertz curve is asymmetric about its point of inflection. The parameters θ_3 and θ_4 control the shape of the function. Below we discuss more properties of the Gompertz curve in detail.

2.3.1.8 JANOSCHEK'S CURVE

The Janoschek growth curve (see Janoschek, 1957) is similar to the Gompertz and Richards curves and reads:

$$y_t = \theta_1 + (\theta_3 - \theta_1) \exp\left(\left[-\frac{t}{\theta_2}\right]^{\theta_5}\right).$$

θ_1 is the asymptotic value again. In contrast to the Gompertz and logistic growth curves, the Janoschek growth curve has a flexible relative inflection ordinate. The Janoschek curve is especially used to model growth of bones.

2.3.1.9 LOGISTIC CURVE

The logistic function (logistic curve) is another popular model of the S-curve representation of growth. The logistic function can be defined by:

$$y_t = a \left(\frac{1 + m \exp(-t/\tau)}{1 + n \exp(-t/\tau)} \right),$$

for real parameters a, m, m, and τ. The logistic function is mostly used to describe the self-limiting growth of a biological population. A popular interpretation is the so-called Verhulst equation. Suppose we have the following differential equation describing growth:

$$\frac{dy}{dt} = \theta_2 y \left(1 - \frac{y}{\theta_1} \right),$$

where θ_2 is the Malthusian growth parameter. Let θ_1 denote the maximum capacity, the solution is:

$$y_t = \frac{\theta_1 \theta_3 \exp(\theta_2 t)}{\theta_1 + \theta_3 (\exp(\theta_2 t) - 1)},$$

where θ_3 is the value of y_t at $t = 0$. A special case is the so-called sigmoid function. Take $a = 1, m = 0, n = 1, \tau = 1$, and we get:

$$y_t = \frac{1}{1 + \exp(-t)},$$

which is the solution to the differential equation $dy/dt = y(1 - y)$ with $y_0 = 1/2$.

A generalization of the logistic function, also known as Richards' curve, reads:

$$y_t = \theta_1 + \frac{\theta_3 - \theta_1}{(1 + \theta_5 \exp[-\theta_2(t - \theta_4)])^{1/\theta_5}},$$

has two horizontal asymptotes: the upper limit θ_3 and the lower limit θ_1. θ_2 is the growth rate again and θ_4 the inflection point. The inflection point has an interesting property in the modelling of world records. According to Nevill and Whyte (2005) this point reveals the period of greatest gain (acceleration) in world record performance. For instance, for running events the inflection point is in the years 1940-1960.

In estimating the model, and assuming $\theta_5 = 1$ it is handy to use the transformation:

$$y_t^* = \log \left(\frac{\theta_2 - y_t}{y_t - \theta_1} \right),$$

and estimate $y_t^* = \theta_3(t - \theta_4)$ to get initial estimates of θ_3 and θ_4.

Another specification of the logistic model is the Brain-Cousens form:

$$y_t = \theta_1 + \frac{\theta_3 + \theta_6 t - \theta_1}{(1 + \theta_5 \exp[-\theta_2(t - \theta_4)])^{1/\theta_5}},$$

where a linear trend $\theta_6 t$ is added in the numerator (see Brain and Cousens, 1989).

2.3.1.10 SCHNUTE'S CURVE

Schnute (1981) proposed a generalization of the Chapman-Richards, Gompertz, and logistic functions:

$$y_t = \left(\theta_3^{\theta_5} + (\theta_1^{\theta_5} - \theta_3^{\theta_5}) \left[\frac{1 - \exp(-\theta_2(t - T_0))}{1 - \exp(-\theta_2(T_N - T_0))} \right] \right)^{\frac{1}{\theta_5}}.$$

Contrary to the Chapman-Richards it does not impose an asymptotic trend. The starting values of the parameters θ_3 and θ_1 are the values of y at the first and the last year of observation, T_0 and T_N respectively.

2.3.2 THEORETICAL SELECTION OF CURVES

Choosing one specific functional form y_t for a set of world records in running is a difficult task, especially because time series are relatively short. In case of modelling world records there is a second complication: some records stand for a very long time, which makes it difficult to observe patterns in improvement. Therefore we rely on two sets of arguments: "theory" and "fit." We start with the former. Wellock, Emmans, and Kyriazakis (2004) provide criteria to select growth functional forms in biology:

1. Fewer parameters are preferred;

2. Functions in which the parameters can be given a theoretical meaning are preferred;

3. Functions with the ability to be expressed in the "rate as a function of the state" are preferred;

4. Biological growth should be seen as a continuous process;

5. Functions with an asymptotic value are preferred;

6. Functions that predict relative growth rate will decrease continuously toward zero, as time evolves, are preferred.

According to these criteria Wellock et al. (2004) argue that the Gompertz, the logistic, and the Von Bertalanffy (Chapman-Richards) are candidate functions. Of course, progress in world records cannot be considered to be a continuous process. And also the theoretical interpretation of the parameters is troublesome, but the other criteria seem to fit the idea of modelling athletic progress. For instance increased professionalism in sports leads to saturation in the sense that recently improvements are generally smaller. This, however, may not be true in sports where technology plays an important role (like in speed skating).

TABLE 2.1
100 meter dash for men: model fit.

Model	SSR	θ_1	θ_2	θ_3	θ_4	Other θs
Linear $\theta_3 - \theta_2 t$	0.607	-	0.009	10.78	-	-
Exponential $\theta_1 + (\theta_3 - \theta_1)\exp(-\theta_2 t)$	0.213	9.554	0.015^a	10.886	-	-
Weibull $\theta_1 + (\theta_3 - \theta_1)\exp(-\theta_2 t^{\theta_5})$	0.231	7.304	0.023	11.015	-	$\theta_5 = 0.61^a$
Hill $\dfrac{\theta_3 \theta_6^{\theta_5} + \theta_1^{\theta_5}}{\theta_6^{\theta_5} + t^{\theta_5}}$	0.514	9.825	-	10.8^a	-	$\theta_5 = 2.5^a$ $\theta_6 = 36.631$
Chapman-Richards $\theta_3 + (\theta_1 - \theta_3)[1 - \exp(-\theta_2 t)]^{\theta_5}$	0.213	9.289	0.015	10.8863	-	$\theta_5 = 0.86^a$
Gompertz $\theta_1 + \theta_3 \exp[-\exp(\theta_2(t - \theta_4))]$	0.217	9.637	0.003	245^a	-	-
Janoschek $\theta_1 + (\theta_3 - \theta_1)\exp([-\frac{t}{\theta_2}]^{\theta_5})$	0.255	8.223	250^a	10.966	-	$\theta_5 = 0.69^a$
Logistic $\theta_1 + \dfrac{\theta_3 - \theta_1}{(1 + \exp[-\theta_2(t - \theta_4)])}$	0.610	1823	0.002	0.011	56	-
Schnute $\left(\theta_3^{\theta_5} + (\theta_1^{\theta_5} - \theta_3^{\theta_5})\left[\frac{1 - \exp(-\theta_2(t - T_0))}{1 - \exp(-\theta_2(T_N - T_0))}\right]\right)^{\frac{1}{\theta_5}}$	0.234	9.788	0.015	10.86^a	-	$\theta_5 = 4.33^a$

^{a} indicates a nonestimated parameter.

2.3.3   FITTING THE MODELS

Next we turn to the fit. We first plot the data (in seconds) against time. This provides valuable information on the shape of the development of world records. For long time series we find a nonlinear S-type pattern. Next we simply fit for one distance, the men's 100 meter dash, various models, compute the sum of squared residuals, and try to interpret the estimated parameters. Table 2.1 gives the results. Note that we did not estimate all the suggested functional forms (e.g., the asymmetric exponential and Schnute equation). The general picture that emerges is that the linear model and the Hill model generally underperform. The other models are comparable in terms of the fit. The Janoschek model gives a rather unreliable estimate of the limit value, as does the logistic model. Based on theory and the fit, the choice seems to be among the Chapman-Richards and the Gompertz models. In this paper we choose the Gompertz model (Gompertz, 1825).

2.3.4   THE GOMPERTZ CURVE IN MORE DETAIL

The four-parameter Gompertz model is given by

$$y_t = \theta_1 + \theta_3 \exp\left(-\exp\left(\theta_2 \left(t - \theta_4\right)\right)\right),$$

where y_t denotes running time in seconds for a certain distances and t is the time index. Four parameters $\theta_1, \theta_2, \theta_3$, and θ_4 are estimated. Parameter θ_1 allows the lower asymptote to be different from zero. Parameters θ_2 and θ_3 control the shape of the curve: for positive values of θ_2 the curve is monotonically decreasing. Parameter θ_2 also determines the smoothness of the curve, small values make the curve linear. The smoothness decreases if θ_2 increases. Parameters θ_1 and θ_4 shift the curve: θ_1 shifts the curve up and down, and θ_4 shifts the curve along the x-axis. Figure 2.1 illustrates this.

The lower and upper limits are identified as:

$$\lim_{t \to \infty} c_t = \theta_1,$$

$$\lim_{t \to -\infty} c_t = \theta_1 + \theta_3,$$

with $\theta_1 > 0$ and $\theta_3 > 0$. The lower limit is calculated from one parameter, whereas the upper limit is the sum of two parameters.

Unlike the logistic model, the Gompertz model is not symmetric about the inflection point (see Schabenberger, Tharp, Kells, and Penner, 1999). Note that in point of inflection (where $t = \theta_4$):

$$y_t = \theta_1 + \frac{\theta_3}{e},$$

where e is the base of the natural logarithms, sometimes called the Eulian number and is defined as

$$e = \lim \left(1 + \frac{1}{n}\right)^n \approx 2.718281828.$$

The point of inflection is below the center of the limits (asymmetry) since $e > 2$:

$$\theta_1 + \frac{\theta_3}{e} < \theta_1 + \frac{\theta_3}{2}.$$

This implies that the Gompertz model has a shorter period of fast growth. At point of inflection the slope is:

$$\frac{dy_t}{dt} = -\frac{\theta_2 \theta_3}{e} < 0,$$

since $\theta_2, \theta_3 > 0$.

The Gompertz function is flexible enough to allow for various patterns. Figure 2.1 shows some examples.

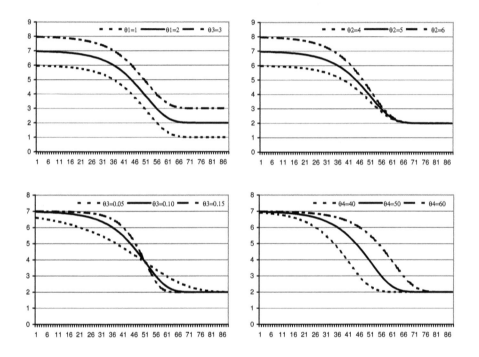

FIGURE 2.1

The shape of the Gompertz function for different parameter values keeping the other parameters constant.

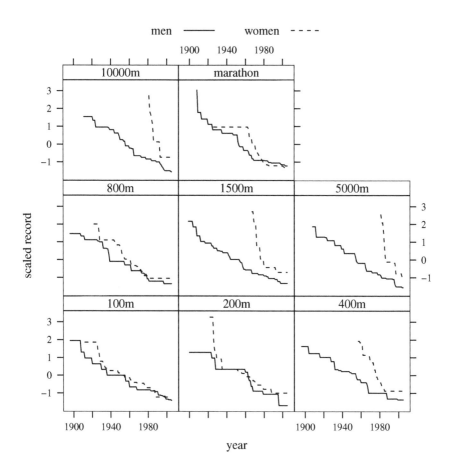

FIGURE 2.2

The development of world records running for men and women (in seconds, normalized). Series are scaled using zero mean and unit standard deviation.

TABLE 2.2
Current running world records (June 2006).

Men	Performance	Athlete	Nat.	Place	Date
100m	9.77	Asafa Powell	JAM	Athens (GRE)	14-06-2005
200m	19.32	Michael Johnson	USA	Atlanta (USA)	01-08-1996
400m	43.18	Michael Johnson	USA	Sevilla (ESP)	26-08-1999
800m	1:41.11	Wilson Kipketer	DEN	Cologne (GER)	24-08-1997
1500m	3:26.00	Hicham El Guerrouj	MAR	Rome (ITA)	14-07-1998
5000m	12:37.35	Kenenisa Bekele	ETH	Hengelo (NED)	31-05-2004
10000m	26:17.53	Kenenisa Bekele	ETH	Brussels (BEL)	26-08-2005
Marathon	2:04:55	Paul Tergat	KEN	Berlin (GER)	28-09-2003

Women	Performance	Athlete	Nat.	Place	Date
100m	10.49	Florence Griffith-Joyner	USA	Indianapolis (USA)	16-07-1988
200m	21.34	Florence Griffith-Joyner	USA	Seoul (KOR)	29-09-1988
400m	47.60	Marita Koch	GDR	Canberra (AUS)	06-10-1985
800m	1:53.28	Jarmila Kratochvílová	TCH	Munich (GER)	26-07-1983
1500m	3:50.46	Yunxia Qu	CHN	Beijing (CHN)	11-09-1993
5000m	14:24.53	Meseret Defar	ETH	New York City (USA)	03-06-2006
10000m	29:31.78	Junxia Wang	CHN	Beijing (CHN)	08-09-1993
Marathon	2:15:25	Paula Radcliffe	GBR	London (GBR)	13-04-2003

2.4 RUNNING DATA

Our main source of information is the official website of the Olympic Movement (www.olympic.org/uk/utilities/reports/) which publishes information on current world records and progress of world records for various events. In this section we show the current record holders and we illustrate the progress of the world records graphically.

For running we analyze distances from the 100 meter sprint to the marathon for both men and women. We transform the raw data of running times to seconds per meter. The time series are shown in Figure 2.2. This figure shows a remarkable improvement in the world record running times. Figure 2.2 indicates that for some womens' distances there are relatively few observations. Also many of the womens' world records are relatively old (see Table 2.2). This makes it difficult to estimate the parameters in the model. Examples of old records are the records on the sprint distances from the late Florence Griffith Joyner, who set the current 100 meter and 200 meter world records in the summer of 1988, almost twenty years ago. The oldest world record in running is the 800 meter record held by the Czech athlete Jarmila Kratochvílová. Also Marita Koch's 400 meter world record (from 1985) is yet to be beaten. The men's world records are improved more often. However, there are examples of world records lasting for a very long time. Jim Hines's 100 meter world record, set in 1968 at the Olympic Games in Mexico was not improved until 1983. Pietro Mennea's 200 meter world record of 1979 lasted until 1996. On May 12, 2006, Justin Gatlin equalled the 100 meters world record of 9.77 seconds set by Asafa Powell in 2005. But Gatlin's record time has been revoked recently following a failed drug test.

Note that the average speed in the men's 200 meters of 10.35 m/s is actually faster than in the men's 100 meters (10.24 m/s). This is due to the fact that it takes time to accelerate. However, it does not hold for the women's world records.

2.5 RESULTS OF FITTING THE GOMPERTZ CURVES

In this section we model the development of the world records. The world records describe the possibility frontier for athletes. We model the trend in world record times. We try to predict the pattern of world record improvements based on past performance assuming that there will not be major technological improvements that shift out the possibility frontier. With only a few observations it is difficult to estimate the shape of the frontier in a reliable way. Despite these problems we will try to determine the frontier, which we use to derive the limits for eight running events — 100m, 200m, 400m, 800m, 1500m, 5000m, 10000m, and the marathon — for both men and women. We use annual data for the world records as they are on December 31 of each year. The last year included in our analysis is 2005 (so Defar's 5000 meter world record is not included in our estimations). To fit the Gompertz model we minimize the sum of squared errors in a grid search in which the parameter θ_2 is fixed, and the remaining parameters are estimated using nonlinear least squares.

TABLE 2.3
Modelling running time (seconds) for the period until 2005 (men).

Men	100m	200m	400m	800m	1500m	5000m	10000m	marathon
θ_1	9.637	18.853	42.999	101.107	195.553	743.602	1575.143	7513.437
(se)	(0.038)	(0.104)	(0.075)	(0.169)	(1.796)	(3.773)	(5.203)	(38.289)
θ_2	236	3	7	16	7840	237	390	3000
θ_3	0.003	0.021	0.026	0.029	0.003	0.019	0.026	0.029
(se)	(0.000)	(0.001)	(0.001)	(0.001)	(0.000)	(0.000)	(0.001)	(0.001)
θ_4	574.733	78.233	54.064	53.079	589.401	54.899	61.415	48.173
(se)	(32.700)	(4.984)	(1.481)	(1.357)	(31.759)	(2.554)	(1.717)	(1.613)
First year	1896	1900	1896	1896	1899	1920	1911	1908
Observations	110	106	110	110	107	86	95	98
Improvements	16	10	16	17	30	25	30	24
R^2	0.979	0.947	0.984	0.979	0.985	0.986	0.983	0.959
SSR	0.217	2.196	5.585	43.606	250.685	1614.260	13380.38	2026859
Limit (sec)	9.55	18.54	42.42	99.01	190.98	733.60	1544.48	7256.36
Limit (h:m:s.s)	9.55	18.54	42.42	1:39.01	3:10.98	12:13.60	25:44.48	2:00:56
2008	9.71	18.94	42.51	1:39:08	3:20.64	12:26.76	25:53.15	2:01:11

TABLE 2.4

Modelling running time (seconds) for the period until 2005 (women).

Women	100m	200m	400m	800m	1500m	5000m	10000m	marathon
θ_1	10.186	21.537	47.556	113.129	230.939	868.397	1767.419	8467.090
(se)	(0.120)	(0.157)	(0.061)	(0.223)	(0.293)	(1.599)	(4.638)	(46.313)
θ_2	3140	5960	7	40	348	30400000	33000000	4800
θ_3	0.003	0.006	0.103	0.047	0.046	0.014	0.014	0.160
(se)	(0.000)	(0.001)	(0.004)	(0.001)	(0.003)	(0.002)	(0.002)	(0.010)
θ_4	−631.603	−290.962	77.902	58.808	50.294	−100.481	−99.334	76.544
(se)	(64.892)	(31.966)	(0.392)	(0.460)	(1.296)	(27.244)	(20.927)	(0.334)
First year	1920	1922	1957	1940	1967	1981	1981	1962
Observations	86	83	49	66	39	25	25	44
Improvements	19	16	12	21	7	8	7	20
R^2	0.950	0.881	0.985	0.985	0.973	0.918	0.953	0.974
SSR	1.764	28.272	3.378	63.220	56.473	386.708	2564.169	2229555
Limit (sec)	9.88	20.30	46.67	110.42	227.87	859.17	1744.23	7982.27
Limit (h:m:s.s)	9.88	20.30	46.67	1:50.42	3:47.87	14:19.17	29:04.23	2:13:02
2008	10.17	20.34	46.67	1:50.42	3:47.87	14:19.27	29:04.97	2:13:02

The regression model is

$$y_t = \theta_1 + \theta_3 \exp\left(-\exp\left(\theta_2\left(t - \theta_4\right)\right)\right) + \eta_t,$$

where η_t is the error term. To find the limit we add the smallest error to parameter θ_1.

$$\text{limit} = \theta_1 + \min \eta_t,$$

The results are in Table 2.3 and Table 2.4. The table gives the parameter estimates with the standard errors (se) between brackets. We also indicate the start of the sample, the number of observations, the number of improvements in the sample, indicators of the fit of the model, and the implied limit values. We also indicate the predicted times for the year of the next Olympiad in Bejing (2008). Especially for the records for women we find a rather flat limit after 2008.

2.6   LIMIT VALUES OF TIME AND DISTANCE

In the previous section we have estimated limit running times for each running distance and for both sexes. This information is shown in Figure 2.6 which indicates that the relationship between the limits of running time and distance is linear in log-log form. This confirms what has been found by Lietzke (1954) and Grubb (1998).

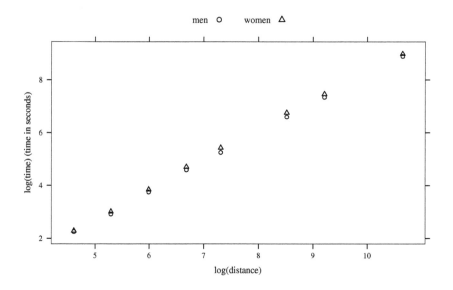

FIGURE 2.3

The log-log relationship between the limits of running time for men and women and distance.

Next, we estimate the pooled log-log model to determine parameter k in Equation (2.1). The model allows for a different intercept for men and women. The null of equal gender intercepts is rejected at 5%. We assume a cross-section SUR specification to allow for contemporaneous correlation between men and women. We also estimate robust coefficient standard errors. In Table 2.5 we applied the White period robust coefficient variance estimator which accommodates arbitrary serial correlation and time-varying variances in the disturbance. Other robust coefficient variance estimators give similar results. Note however that the sample is very small.

TABLE 2.5
Cross-section SUR estimation results for the log-log relationship between limit time t and distance d, with the White period robust coefficient variance estimator.

Model: $\log t = k \log d + \log a$		
Parameter	Coefficient	Robust standard error
a	−2.8058	0.0697
k	1.1067	0.0082
Gender-specific intercepts		
a-Men	−0.0551	
a-Women	0.0551	
R^2	0.9998	
s.e. of regression	1.0859	
Durbin-Watson	1.4669	
Redundant FE $F(1, 13)$ (p-value)	48.8978	(0.0000)

Our estimate for k is slightly smaller than the one Kennelly reported in 1905. Kennelly's estimate is 9/8, while our estimate is 11/10, which is significantly smaller than 9/8 at 5%. This implies that if distance increases by 10%, the limit time increases by 11%, irrespective of gender. This is what Lietzke refers to as the exhaustion constant or the fatigue rate. The gender difference of 2×0.0551 is significant. The estimates imply that the women are 11% slower than men.

Finally, we estimate the limit time for the one-mile run (1609 meters), which is an irregular running distance at main championships. The current world record is held by Hicham El Guerrouj (MOR) who set a world record time of 3:43.13 on July 7, 1999 in Rome (ITA). The current women's world record on the mile is 4:12.56, set by Russian Svetlana Masterkova on August 14, 1996 in Zürich (SUI). The development of the one-mile world records are shown in Figure 2.4.

The estimates in Table 2.5 imply a one-mile limit time for men of 3:22.48 (202.48 seconds). The estimated women's one-mile limit is well below the 4 minute barrier at 3:46.04 (226.04 seconds). The question is how these one-mile limit times differ

TABLE 2.6
Modelling running time for the one-mile run (seconds).

	Men		Women	
	Coefficient	Std. error	Coefficient	Std. error
θ_1	223.3653	0.2115	253.2729	0.4123
θ_2	39		57	
θ_3	0.0330	0.0004	0.0672	0.0034
θ_4	91.2639	0.6382	108.6841	0.4200
First year	1896		1967	
Observations	110		39	
Improvements	35		10	
R^2	0.9944		0.9730	
SSR	82.4707		74.043	
Limit (sec)	220.97		250.73	
Limit(m:s.s)	3:40.97		4:10.73	

from separate estimates of the Gompertz curve for the one-mile run. Two matters complicate this exercise. First, the current world records are rather old. Second, the one-mile run is an irregular running distance. Table 2.6 shows the results.

The Gompertz model suggests limit world records which are considerably slower than the times implied by Table 2.5. If the one-mile run was a regular running event, the limit world record would be 18.5 seconds faster for men and 24.7 seconds for women.

2.7   SUMMARY AND CONCLUSIONS

In this paper we analyze the development of world records. We focus on running events and fit the historical curves of individual events using an asymmetric S-shape Gompertz curve. We compute the implied infinite lower bounds for each event and for men and women. With only a few observations for some events, it is difficult to estimate the shape of the frontier in a reliable way. Despite these problems we have determined this frontier, which we use to derive the limits for eight running events — 100, 200, 400, 800, 1500, 5000, 10000 meter, and the marathon — for both men and women. We use annual data for the world records until 2005.

The lower bounds calculated from the Gompertz curve estimations are compared in a cross-sectional setting to arrive at a relationship between time and distance. Our estimate is slightly different from the one Kennelly reported in 1905. Our results imply that if distance increases by 10%, the limit time increases by 11%, irrespective of gender.

The gender difference is significant. The estimates imply that women are 11% slower than men. Finally, we forecast the lower bound of the mile using the log-log specification between time and distance. If the one-mile run was a regular running

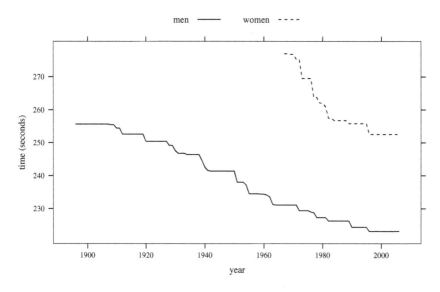

FIGURE 2.4
The development of one-mile world records for men and women (in seconds).

event like the other events, the gain in the long run would be 18.5 seconds for men and 24.7 seconds for women.

REFERENCES

von Bertalanffy, L. (1938). A quantitative theory of organic growth. *Human Biology 10*, 181–213.

Blest, D.C. (1996). Lower bounds for athletic performances. *The Statistician 45*, 243–253.

Brain, P. and R. Cousens (1989). An equation to describe dose responses where there is stimulation of growth at low dose. *Weed Research 29*, 93–96.

Charnes, A., W.W. Cooper, and E. Rhodes (1978). Measuring the efficiency of decision making units. *European Journal of Operational Research 2*, 429–444.

Chatterjee, S. and S. Chatterjee (1982). New lamps for old: An exploratory analysis of running times in Olympic Games. *Applied Statistics 31*, 14–22.

Deakin, M.A.B. (1967). Estimating bounds on athletic performance. *Mathematics Gazette 51*, 100–103.

Francis, A.W. (1943). Running records. *Science 98*, 315–316.

Gompertz, B. (1825). On the nature of the function expressive of the law of human mortality, and on a new mode of determining life contingencies. *Philosophical Transactions of the Royal Society of London 115*, 513–585.

Grubb, H.J. (1998). Models for comparing athletic performances. *The Statistician 47*, 509–521.

Hill, A.V. (1913). The combinations of haemoglobin with oxygen and with carbon monoxide. *Biochemistry 7*, 471–480.

Janoschek, A. (1957). Das reaktionskinetische Grundgesetz und seine Beziehungen zum Wachstums- und Ertragsgesetz. *Statistische Vierteljahresschrift 10*, 25–37.

Kennelly, A.E. (1905). *A study of racing animals*. The American Academy of Arts and Sciences.

Kuper, G.H. and E. Sterken (2003). Endurance in speed skating: The development of world records. *European Journal of Operational Research 148*(2), 293–301.

Lietzke, M.H. (1954). An analytical study of world and Olympic racing records. *Science 119*, 333–336.

Malthus, T.R. (1798). *An Essay on the Principle of Population*. London: J. Johnson. Available at http://www.econlib.org/Library/Malthus/malPop.html.

Mosteller, F. and J.W. Tukey (1977). *Data Analysis and Regression*. Reading, MA.: Addison-Wesley.

Nevill, A.M and G. Whyte (2005). Are there limits to running world records? *Medicine and Science in Sports and Exercise 37*, 1785–1788.

Ratkowsky, D.A. (1990). *Handbook of Nonlinear Regression Models*. New York: Marcel Dekker.

Richards, F.J. (1959). A flexible growth function for empirical use. *Journal of Experimental Botany 10*, 290–300.

Schabenberger, O., B.E. Tharp, J.J. Kells, and D. Penner (1999). Statistical tests for Hormesis and effective dosages in herbicide dose response. *Agronomy Journal 91*, 713–721.

Schnute, J. (1981). A versatile growth model with statistically stable parameters. *Canadian Journal of Fishery and Aquatic Science 38*, 1128–1140.

Smith, R.L. (1988). Forecasting records by maximum likelihood. *Journal of the American Statistical Association 83*, 331–388.

Sterken, E. (2005). A stochastic frontier approach to running performance. *IMA Journal of Management Mathematics 16*, 141–149.

Tatem, A.J., C.A. Guerra, P.M. Atkinson, and S.I. Hay (2004). Momentous sprint at the 2156 Olympics? *Nature 431*, 525.

Wainer, H., C. Njue, and S. Palmer (2000). Assessing time trends in sex differences in swimming and running. *Chance 13(1)*, 10–21.

Wellock, I.J., G.C. Emmans, and I. Kyriazakis (2004). Describing and predicting potential growth in the pig. *Animal Science 78*, 379–388.

Whipp, B.J. and S.A. Ward (1992). Will women soon outrun men? *Nature 355*, 25.

3

The physics and evolution of Olympic winning performances

Ray Stefani

California State University

ABSTRACT

The physical laws that govern the motion of a runner, jumper, swimmer, rower, and speed skater provide an understanding of the amount of power required to master each sport and a list of factors that define success in each sport. There have been four noteworthy technical breakthroughs: the rowing ergometer, the Fosbury Flop technique for the high jump, clap skates used in speed skating, and the fiberglass pole now used in the pole vault. Percent improvement per Olympiad, designated $\%I/O$ was used to evaluate the rate of improvement. For all events in this study, the average $\%I/O$ was 0.70% for running, 1.56% for jumping, 1.54% for swimming, 1.25% for rowing, and 1.62% for speed skating. The events with more technical challenges thus exhibited about twice the improvement of running. Athletes have often been held captive to outside influences beyond the control of sport, as in the negative influences of WW1, WW2 and boycotts, the positive influence of the cold war period, while recently, certain event performances have become worse after significant efforts to punish doping were initiated. Long-term improvement has been gradually declining in running, jumping, swimming, and rowing, while speed skating has retained consistent improvement. Women who had been improving faster than men in running, jumping, and swimming events are now improving at about the same rate.

3.1 INTRODUCTION

A casual review of a sports almanac or today's news reveals a montage of records being broken, technological improvements emerging, political events affecting sports and, unfortunately, the use of performance-enhancing drugs. We wonder where performances are headed. The purpose of this section is to examine certain sports events in a holistic way so as to arrange the montage of facts into a coherent picture. We shall examine how the laws of physics provide insight into those factors by which performances may be improved. We shall examine the underlying political events and technological breakthroughs that strongly influenced performances.

Which events and performances shall we examine? If we formulate a taxonomy of sports events (Stefani, 1999), we find two important descriptors: the evaluation

method by which a winner is determined and the manner of interaction between an
athlete and the opponent. The winning athlete may be evaluated by subjective judging (as in boxing and diving), by arbitrary scoring (as in basketball and shooting),
and by unambiguous measurement (as in swimming and running). Competitors interact in the only three ways that two separate objects can interact: by direct, indirect,
and independent movement. Competitors interact directly in what may be termed a
combat sport, wherein the goal is to control the opponent as in boxing and wrestling.
Competitors interact indirectly, in what can be termed an object sport, in that the
competitor tries to control some object as in basketball and soccer. Competitors may
only have incidental contact in what may be termed an independent sport, such as
swimming and rowing, wherein the competitor tries to control the competitor's own
self to succeed. The performance of an athlete in an independent sport is not directly
influenced by the opponent as in a combat sport and an object sport. Further, those
sports decided by unambiguous measurement provide a better comparison over time
than those sports decided by subjective judging. This section focuses on those sports
in which the competitor is evaluated by unambiguous measurement in independent
events. We shall therefore examine running, jumping, swimming, rowing, and speed
skating events.

What performance data shall we examine? World records provide a limited amount
of evolutionary information. World records are set under ideal conditions, at irregular
intervals and are, by definition, always improving (see for example Chapter 2 of this
volume). No information is provided when there may be a downturn in performance
due to warfare, for example. We shall examine Olympic winning performances for
the sports just mentioned, because Olympic competition occurs at regular intervals
and winning performances represent the state-of-the-art as of that time. Those winning performances provide a picture into the past and perspective for predicting the
future.

3.2   RUNNING EVENTS

3.2.1   THE PHYSICS OF RUNNING

An athlete is subject to the same laws of physics that govern the motion of an au-
tomobile, boat, and airplane. Those laws provide understanding of the basic factors
that an athlete may employ to improve performance. The output power of an athlete will be examined here. Internal metabolic power is not considered. That power
would be about four times the output power, since the body is an inefficient user of
metabolic energy. According to the laws of physics, power is defined (Lerner, 1996)
as the rate of change of energy and, equivalently, as the product of force times the
component of velocity in the direction to which force is applied. Power is measured
in watts, much as for a light bulb, time is in seconds (s), velocity is in meters/second
(m/s), while cleared height, cleared length, and race distance are all in meters (m).
As an athlete accelerates from walking to running, the athlete's center of gravity,
abbreviated CG, begins to follow a curved trajectory similar to that of a thrown object. At the moment the athlete's feet both leave the ground, the velocities are as in

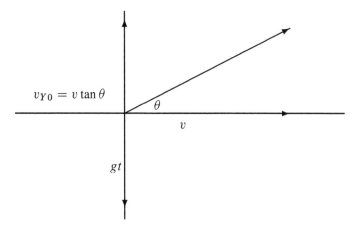

FIGURE 3.1
Horizontal and vertical velocities for running and jumping.

Figure 3.1 for time $t = 0$. At that moment, the CG moves with a resultant velocity v_R, making an angle θ with the horizontal. The horizontal component is v, and the vertical component is v_{Y0}. At these relatively low running velocities, we can ignore aerodynamic effects, so that the horizontal velocity remains constant until the foot again touches the ground. During this ballistic phase, the only applied force is due to the downward acceleration of gravity g, equal to 9.8m/s². The resultant vertical velocity v_Y is then equal to $v_{Y0} - gt$. The running athlete's CG follows the trajectory of Figure 3.2 covering the range L. Other parts of Figure 3.2 will be covered in the context of jumping events. At the highest point, when the CG has risen by d meters, the vertical velocity is zero which happens in v_{Y0}/g seconds. At that moment the kinetic energy has changed from an initial value of $(1/2)mv_Y^2 = (1/2)mv_{Y0}^2$ to $(1/2)mv_Y^2 = 0$, where m is the mass of the athlete measured in kg. The letter "m" is used in two contexts: as an abbreviation for the unit of length, denote by standard typed "m" and as an algebraic symbol for mass, denoted by italic typed "*m*." The power applied to the CG, P_{CG}, is equal to the energy change divided by time:

$$P_{CG} = \frac{(1/2)mv_{Y0}^2}{v_{Y0}/g} = mgv_{Y0}/2. \tag{3.1}$$

Because of inefficient running style and energy lost in ground contact, not all of the athlete's output power P results in P_{CG}. Let efficiency be defined by e, which varies from 0 to 1. Then Pe equals P_{CG} causing $P = P_{CG}/e$. Finally, P will be expressed in terms of the horizontal velocity v, which can be found from elapsed running time. From Figure 3.1, v_{Y0} equals $v \tan \theta$. Thus, the athlete's output power is given by

$$P = \frac{mgv \tan \theta}{2e}. \tag{3.2}$$

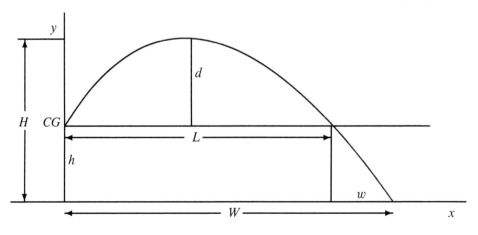

FIGURE 3.2
Trajectory of the *CG* for the velocities of Figure 3.1.

This exercise in applied physics allows us to list factors that may be used to improve running time. The mechanics of running that create values of P are such that a runner achieves increased velocity by creating increased ground reaction force (Weyand, Sternlight, Bellizzi, and Wright, 2000) and not by faster leg movement. The sweep time for an athlete's noncontact leg is about the same for athletes at differing velocities. An increase in ground reaction force causes the leg to sweep farther forward during the same sweep time, generating a longer stride and greater velocity. Further, data gathered by Maud and Schultz (1986) on a cycle ergometer, by Lutoslawska, Klusiewics, Sitkowsi, and Krawczyk (1996) on a rowing ergometer and by Iwaoka, Hatta, Atomi, and Miyashita (1988) on a treadmill indicate that power output depends on the level of training of an athlete and on the athlete's lean body mass. The factors listed in Table 3.1 affect the output power of a runner. These same factors must be improved over time to improve performance.

England's Harold Abrahams, featured in the movie *Chariots of Fire*, won the 100m in 1924 in a time of 10.6s. The long-term improvement in running times is exemplified by the fact that a relay of two Harold Abrahams each running 10.6s would not have won the individual Olympic 200m in 2004, since the winning time was 19.79s. The 100m was won in 2004 in 9.85s. To be fair to Abrahams, Equation 3.2 allows us to account for changes in track surface rebound tendency and track shoe resiliency. To simplify the analysis, let the runners' body masses be 80kg and let $\tan\theta$ equal 0.14, equivalent to a running angle of about 8°, for a reasonably smooth stride. Due to the stiff track shoe of 1924 and soft cinder track, let the relative efficiency e be 0.95 in 1924. Using Equation 3.2, Abrahams power output for the 100m run would be $80\times9.8\times(100/10.6)\times0.14/(2\times0.95)$ or 545.0W. The gold medal winner in 2004

TABLE 3.1
Factors influencing the power output of a runner.

Physiology	
	Size and general fitness of the pool of athletes
	Lean body mass
	Muscle specific training
Techniques	
	Foot-track contact
	Vertical motion/smooth stride
Coaching the technique	
	Record past performances
	Evaluate performances
	Train for proper technique
Quality of the venue and personal equipment	
	Shoe resiliency
	Track surface characteristics

produced 557.2W, assuming $e = 1$. If Abrahams could have produced 545.0W with a relative efficiency of 1 compared to today's athlete, then his Equation 3.2 would be multiplied by e and his velocity would be divided by e, which would cause his time to be multiplied by e, for an elapsed time of 10.6×0.95 or 10.07s, good enough to win a gold medal in 1980.

3.2.2 MEASURING THE RATE OF IMPROVEMENT IN RUNNING

Based on more than 30 years of research, I have found it helpful to use a dimensionless measure of the rate of improvement, one that can be used for events contested under widely differing time scales. Let $w(i, n)$ denote the winning time for event i contested during Olympics n. I define the percent improvement per Olympiad for event i during Olympics n as

$$\%I/O = \%I/O(i, n) = 100 \times \left(1 - \frac{w(i, n)}{w(i, n - 1)}\right). \tag{3.3}$$

For example, the winning time for the men's 400m run was 49.3s in 1960 and 50.1s in 1956. The $\%I/O(400m, 1960)$ would be $100 \times (1 - 49.3/50.1)$ or 1.60%. When there is a span of inactivity (as during war time) the intervening period is defined as one Olympiad for convenience. If Equation 3.3 is applied to the winning performances of past Olympic running events, then we can find the average rate of improvement for a set of events over a set of Olympic competitions. For example we might want the average $\%I/O$ for men in all running events contested in 2006 or the average $\%I/O$ for women in all running events contested during the cold-war period. Walking and marathon events are not included because the course elevations and running surfaces are not constant for each of the various competitions.

3.2.3   PERIODS OF SUMMER OLYMPIC HISTORY

The modern Summer Olympic span of time from 1896 to 2004 may be divided into five periods, in order to examine rates of improvement in the context of the historical events of each era. These five periods include the seven Summer Olympic Games spanning WW1 (1896-1924), from the start of the modern Games until the second Games after WW1, the five Games similarly spanning WW2 (1928-1952), the six Games spanning the cold war (1956-1976), the two boycotted Games plus one fully attended Games (1980-1988) and the four recent Games (1992-2004). Three Games were not celebrated due to war: the Games of 1916, 1940, and 1948. Table 3.2 contains $\%I/O$ values for running events, averaged separately for men and women for each Olympics and then averaged at the bottom of the table for each of the five periods. Jumping and swimming results will be discussed later. Recall that a value of $\%I/O$ is always taken for one Olympic winning performance, compared to the immediately preceding Olympics. Thus, the earliest $\%I/O$ value is calculable with the second competition of a given event. Running competition began for men in 1896 and for women in 1928. The number of $\%I/O$ values in running events vary for men from five in 1900 to 12 in 2004 and for women from two in 1932 to 11 in 2004.

For the men's running events in Table 3.2, the highest improvement was in 1900, when the Olympic Games dramatically expanded the pool of athletes after the fledgling start in 1896. Men's $\%I/O$ values followed a similar pattern during the periods spanning WW1 and WW2. Improvement was relatively high during the pre-WW1 build up in 1912, dropped due to the debilitating effects of WW1 in 1920 and then rebounded in the second post-war Games in 1924. Those 1924 Games were dramatized in the movie *Chariots of Fire*. $\%I/O$ for men peaked again during the build-up prior to WW2 in 1932 and 1936, dropped to 0.35% due to war-time effects in 1948 and rebounded in the second post-war Games of 1952. Improvement for men remained consistently high during the cold war Games from 1956-1976 as Eastern-bloc and Western-bloc nations placed a high priority on athletic dominance. Improvement was negative in 1980 when Western-bloc men were absent in Moscow while the value was positive compared to Moscow 1984 when Eastern-bloc men were absent in Los Angeles. Together, the two blocs of athletes pushed each other to greater improvement than when competing separately. Improvement rebounded at the fully attended Games of 1988.

As with men, women's highest running $\%I/O$ value was in their second Olympic competition in 1932. As for men, improvement dropped in the first post-WW2 Games in 1948, rebounded in 1952 and then remained consistent during the cold war. Improvement remained high for women at the Western-boycotted Moscow Games of 1980 but dropped to a negative value at the Eastern-boycotted Los Angeles Games of 1984, since Eastern-bloc women dominated Olympic running medals in that era. As with the men, improvement rebounded at Seoul in 1988.

Improvement for men and women both dropped in the 1992-2000 period, when four of six $\%I/O$ averages were negative, indicating generally declining performance. If we evaluate all athletics (track and field) performances, more than two-thirds (23 of 34) of all comparable winning performances in 1988 would have won at Sydney in

TABLE 3.2
Percent improvement per Olympiad for running, jumping, and swimming.

Olympics	Running %I/O Men	Running %I/O Women	Jumping %I/O Men	Jumping %I/O Women	Swimming %I/O Men	Swimming %I/O Women
1900	9.41		5.90			
1904	1.69		0.55			
1908	0.23		4.34			
1912	1.74		2.17		3.96	
1920	−1.16		−0.91		1.82	11.07
1924	2.55		2.43		4.69	2.87
1928	−0.02		1.55		2.82	4.43
1932	1.01	2.78	1.56	4.21	2.73	3.97
1936	1.38	1.68	2.85	−3.44	1.64	0.96
1948	0.35	−2.38	−2.59	5.00	0.49	2.73
1952	1.46	3.20	2.74	4.49	2.51	1.21
1956	0.88	1.44	2.09	3.58	2.29	3.32
1960	1.12	0.60	2.87	2.71	1.46	2.57
1964	0.49	1.28	2.27	4.41	3.29	2.30
1968	0.73	1.28	4.06	−1.66	1.33	1.43
1972	0.61	0.96	−0.06	2.45	3.51	2.24
1976	0.41	0.74	0.49	0.19	2.70	4.45
1980	−0.68	1.36	3.20	3.18	−0.58	1.15
1984	1.47	−0.75	−0.42	0.56	1.70	−0.16
1988	0.39	2.04	2.01	3.41	0.77	1.26
1992	−0.27	−1.30	−0.19	−2.00	0.58	0.41
1996	0.81	−0.41	0.45	0.61	−0.10	−0.26
2000	−1.10	0.32	−0.88	−1.54	1.15	1.40
2004	0.57	0.24	0.55	2.76	0.64	−0.20
1900-24	1.91		2.41		3.49	6.97
1928-52	0.83	1.53	1.22	2.95	2.04	2.66
1956-76	0.71	1.01	1.95	1.95	2.43	2.72
1980-88	0.39	0.97	1.59	2.38	0.63	0.75
1992-04	0.00	−0.28	−0.02	0.33	0.57	0.34

2000. At www.wada-ama.org, under the heading "Brief History of Anti-Doping," WADA states, "There is an evident connection between more effective test methods and the remarkable drop in the level of top results in some sports in the 1990s." It can be argued that the notoriety of Ben Johnson's disqualification in 1988, the widely-published Canadian Dubin Inquiry (Dubin, 1990), finding widespread use of performance enhancing drugs in track and field and the subsequent increased vigilance of drug testing were significant causes of the low values of $\%I/O$ in running from 1992-2000. $\%I/O$ rebounded somewhat in 2004.

As just discussed, in some Games, $\%I/O$ has been strongly influenced by outside causes such as warfare, boycotting, and drug testing. Averages taken over each of the five periods of Olympic history tend to cancel short-term influences to reveal long-term trends in running performance. The rate of improvement has been gradually declining. Women were improving faster than men; but are now improving at about the same rate. Table 3.3 contains athletes having the three highest $\%I/O$ values in running as well as for other events to follow.

3.2.4 THE FUTURE OF RUNNING

The past teaches us that outside influences such as wars, boycotts, and enhanced drug testing has had as much influence on the rate of improvement in running events as did the sport-related factors listed earlier. We cannot know what the outside influences will be, but we can suggest what will happen regarding the sport-related factors. The rate of improvement will likely continue to lessen as a diminishing return occurs. Further, the primary technical means of improvement is foot-ground contact which improves slowly compared to the events we will examine shortly. If it is assumed that the use of performance-enhancing drugs will at least not increase, then "clean" athletes will begin to be compared to similarly-clean athletes of each earlier Olympics with a positive rate of improvement near the rate seen in 2004.

Women were improving faster than men; but are now improving at about the same rate. I note in Stefani (2006), that the typical trained female elite athlete has 28% less lean body mass than the male counterpart, that researchers have shown that power outputs for equally trained athletes are in proportion to lean body mass and that the application of physics to a variety of recent Olympic winning performances results in that same 28% difference. I suggest that currently, men and women improve at the same rate with a 28% power difference because they are about equally trained. The greater rate of improvement in the past may well have been a "catch up" phase due to improving training opportunities for women.

3.3 JUMPING EVENTS

3.3.1 THE PHYSICS OF JUMPING

High jump and pole vault performances are measured by the vertical height cleared, H in Figure 3.2, which is equal to the starting height h of the CG above the ground plus the increase by d when the athlete performs a jump. That is, $H = d + h$. It is possible to derive power in terms of $d = H - h$. At the height of the jump, the

TABLE 3.3
Best rates of improvement (the second competition in each event was excluded).

%I/O	Winner	Nation	Event	Olympics
		Running		
6.17	Lasse Virin	Finland	M 10km run	1972
5.15	Kip Keino	Kenya	M 3000m steeplechase	1972
4.63	Eddie Tolan	USA	M 100m run	1932
		Jumping		
10.29	Bob Beamon	USA	M Long jump	1968
8.51	Bob Seagren	USA	M Pole vault	1964
7.84	Wolfgang Nordwig	E. Germ.	M Pole vault	1972
		Swimming		
10.17	Andrew Charlton	Australia	M 1500m freestyle	1924
7.39	Warren Kealoha	USA	M 100m backstroke	1920
7.19	Dawn Fraser	Australia	W 100 m freestyle	1956
		Rowing		
11.80	Jährling, Ulrich, Spohr	E. Germ.	M pairs with cox	1980
11.29	Dreifke and Kröppelien	E. Germ.	M double sculls	1980
11.23	Boreyko and Golovanov	USSR	M pairs without cox	1960
		Speed Skating		
8.40	Yevgeny Grishin, Yuri Mikhalov	USSR	M 1500m (tie)	1956
6.94	Yevgeny Grishin	USSR	M 500m	1956
6.40	Thomas Gustafson	Sweden	M 5000m	1988

initial kinetic energy $(1/2)mv_{Y0}^2$ is replaced by potential energy mgd. By finding v_{Y0} (equal to $v \tan \theta$) in terms of d, Equation 3.2 becomes

$$P = \frac{mg^{3/2}\sqrt{(H-h)}}{e\sqrt{2}} \qquad (3.4)$$

What can be learned in terms of the proper techniques that will result in acceptable values of P? Analyses of the high jump by Jacoby and Fraley (1995) and Linthorne (1999) suggest that an athlete should reduce horizontal velocity somewhat for a given take off angle, to achieve a higher vertical velocity component, v_{yo}, which results in greater height cleared. Thus, horizontal velocity, take off angle and take off velocity must be tuned. Those authors suggest that the takeoff angle should be about 50°. It is interesting that the Fosbury Flop technique, which improved high jump performance, and the clap skate, which improved speed skating performance, both employ about the same take off angle and both innovate by allowing greater use of extensor biomechanics.

In the pole vault (McGinnis, 1991, 1997), the flexibility of the pole limits the v that can be directed vertically when the pole is flexed so as to rebound. A pole vault athlete attempts to convert as much as possible of v into v_{yo}, by timing the placement of the pole, hand locations, and body positioning. A fiberglass pole allows conversion of a greater v (and higher resulting d and H) than the stiffer steel and bamboo poles of the past.

Based on Croskey, Dawson, Luessen, Marohn, and Wright (1922) and Hellebrandt and Franssen (1943), the CG of a typical man is located at 56% of height, and the CG of a typical woman is located at 55% of height. Using those values and height data in McArdle, Katch, and Katch (1981), a male jumper has a CG that is 1m above the ground when upright and the similar figure for women is 0.93m. In the pole vault, the athlete plants the pole while in a fairly upright stance; hence h in Equation 3.4 would be about 1m for men and 0.93m for women. Video analyses of world-class high jumpers suggest that an elite high jumper should dip somewhat at take off (Jacoby and Fraley, 1995; Linthorne, 1999) so the CG that is above ground prior to the jump is about 46% of the athlete's height. It follows that h would be 0.84m for men and 0.78m for women. If an 80kg male athlete clears 2.1m (6' 11") in the high jump with $e = 1$, the required power is 1948W. For a similar 6m (19' 8") clearance in the pole vault, P is 3880W. In terms of initial upward velocity, the analysis predicts that a male pole vault athlete should clear $H = d + h = (v_{yo}^2/2g) + 1$m. If a forward velocity of 10m/s is directed upward, the height cleared would be 6.1m (20'), agreeing with current world class performance and approach speed.

As to the physics of the long jump, let L in Figure 3.2 represent the horizontal distance traveled in twice the time for d to be achieved, that is, for the CG to return to the same height as on take off. If v is constant, then L is given by $2(v_{yo}/g)v$. The long jump performance W is equal to L plus a small distance w as the CG ends up at ground level. Equation 3.2 can be written in terms of $L = W - w$, resulting in

$$P = \frac{mg^{3/2}\sqrt{\tan\theta}\sqrt{(W-w)}}{2^{3/2}e}. \tag{3.5}$$

The long jump take off angle (Linthorne, 1999; Wakai and Linthorne, 2000) is about 20° to provide good distance achieved. An analysis was made of several typical v values, each with a takeoff angle 20° and typical CG heights above the ground for men and women. The result suggests that w is approximately 2.1m for men and 1.9m for women. For an 80kg man with $e = 1$ to achieve a long jump of 8.6m (28'10"), the required power would be 1334W.

The output power required by jumping is several times higher than for running, because a much greater kinetic energy is converted to potential energy to create upward movement against gravity and because the critical period when the athlete takes off covers less than one second. The result is that jumping is far more technically challenging. The greater reliance on technique implies more opportunity for improvement of that technique, compared to running, and a greater possibility of a breakthrough performance.

TABLE 3.4
Factors influencing the power output of a jumper.

Physiology	
	Size and general fitness of the pool of athletes
	Lean body mass
	Muscle specific training
Techniques	
	Horizontal velocity
	Take off angle
	Take off velocity
	Position in the air
	Foot-track contact
	Pole vault pole carry and placement
	Pole vault hand positioning
	Pole vault upper body action
Coaching the technique	
	Record past performances
	Evaluate performances
	Train for proper technique
Quality of the venue and personal equipment	
	Shoe resiliency
	Track surface characteristics
	Landing pit quality
	Pole stiffness

3.3.2 MEASURING THE RATE OF IMPROVEMENT IN JUMPING

Improvement in jumping implies a higher value for $w(i, n)$ compared to $w(i, n-1)$, the opposite of the situation for a timed event such as running. $\%I/O$ becomes

$$\%I/O(i, n) = 100 \times \left(\frac{w(i, n)}{w(i, n-1)} - 1 \right). \tag{3.6}$$

For example, the winning men's pole vault was 4.7m (15' 5") in 1960 and 5.1m (16' 8.75") in 1964, the first year that a fiberglass pole was used. For 1964, $\%I/O$ (pole vault, 1964) is $100 \times (5.1/4.7 - 1)$ or 8.51%.

Table 3.2 contains $\%I/O$ values for jumping events, averaged separately for men and women for each Olympics and then averaged at the bottom of the table for each of the five periods. Jumping competition began for men in 1896 and for women in 1928. The number of $\%I/O$ values in jumping events is four for men from 1900 to 2004. The number of values for women varies from two in 1932 to 4 in 2004. The pattern of improvement in jumping is similar to that for running during the first three periods and during the recent period, but values are generally higher than for running, consistent with comments made earlier. The highest $\%I/O$ value for men occurred in the second competition in 1900. Men's jumping showed a negative improvement for the first Games after WW1 and WW2, followed by a rebound in the second Games after each war. Improvement remained consistent during the cold war period. Women's performance was not negatively impacted by WW2, although there were only two such events. Improvement remained high for men and women in 1980 when Western-bloc athletes were absent, since Eastern-bloc athletes dominated those events at the time. Conversely, the absence of Eastern-bloc athletes in 1984 caused a negative improvement for men and a smaller-than-normal improvement for women. I believe that the negative average $\%I/O$ values in 1992 and 2000 were caused by increased vigilance in drug testing as with running events. Improvement has rebounded as with running in 2004 for the same reasons as mentioned for running. Over the five period of this study, $\%I/O$ has gradually diminished and women have improved faster than men prior to the recent period in which the improvement of both genders is not significantly different.

Table 3.3 contains the athletes who achieved the three highest jumping $\%I/O$ values, excluding the second competition in each event. In 1968, Bob Beamon improved the long jump winning performance by 10.29% compared to 1964. Beamon's 1968 performance still held the Olympic record as of the 2004 Games, 36 years later. The next oldest track and field Olympic record is from 1980. His performance has crept into the vernacular in that it is said that a significant breakthrough performance *Beamonizes* a given record.

Seagren and Nordwig were early users of the fiberglass pole. It is possible to use values of $\%I/O$ in Table 3.5 to compare four technological improvements, including the fiberglass pole, that have made significant impacts on Olympic winning performances. For each, the average $\%I/O$ for the event or class of events is calculated from 1956 to the Olympics preceding introduction. That baseline is compared to the first value after introduction. The fiberglass pole increased men's pole vault improvement by 419% when introduced in 1964. The Fosbury Flop increased improvement for men and women high jumpers by a combined average of 83% when introduced for men in 1968 and for women in 1972.

3.3.3   THE FUTURE OF JUMPING

Similar to running, there has been a diminishing return to the improvement of jumping events caused by improving the various factors that were listed earlier. When drug testing has taken its toll on cheaters, future performances of clean athletes should improve as compared to earlier clean athletes as appeared to happen in 2004, although such improvement is likely to occur at a lower level than in the past. Because there are so many factors to work with, jumping events should continue to exhibit about twice the rate of improvement of running events, absent draconian political events to the contrary. Women are likely to improve at about the same rate as their male counterparts. The relative biomechanical complexity of jumping suggests that some future Olympic jumper will *Beamonize* a jumping record.

TABLE 3.5
Four technological breakthroughs.

Technological Breakthrough	Event	Olympics of First Use	Average %I/O for 1956 up to First Use	%I/O Upon First Use	Percent Increase in %I/O
Rowing Ergometer	All Rowing Events	1980	1.22	7.42	508
Fiberglass Pole	Pole Vault	1964	1.64	8.51	419
Fosbury Flop	High Jump	1968 (Men) 1972 (Women)	2.25	4.12	83
Clap Skate	All Speed Skating Events	1998	1.82	2.88	58

3.4   SWIMMING EVENTS

3.4.1   THE PHYSICS OF SWIMMING

In swimming, the downward gravitational force (weight) mg is balanced by an upward buoyancy force; therefore, we do not have to consider up and down motion as we did for running and jumping. Instead, we focus on horizontal motion. In the diagram below, a swimmer applies a force F which propels the swimmer to the right with velocity v, while an equal hydrodynamic drag force pulls the swimmer to the left. An arrow shows the direction of each force.

Drag Force$(\rho, v^2, S, C_d/2)$ ← Swimmer → Force F caused by the swimmer
(Swimmer has velocity $v \rightarrow$)

According to the laws of physics in this case, (Lerner, 1996), the power acting on the CG is Fv. To obtain power output by the swimmer, we must include efficiency e, so that P is Fv/e or

$$P = \frac{\rho v^3 S C_d}{2e}. \tag{3.7}$$

In Equation 3.7, ρ is the density of water, 1000kg/m3, S is the surface area m^2 in contact with the water, and C_d is the dimensionless drag coefficient which accounts for affect of the shape of the swimmer upon drag force. Note that power in swimming depends on the third power of horizontal velocity, while for running and jumping, power depends on the first power of horizontal velocity. The hydrodynamics of swimming have been extensively studied (Toussaint, De Groot, Savelberg, Vervoorn, Hollander, and Van Ingen Schenau, 1988; Toussaint, Janssen, and Kluft, 1991; Toussaint, Knops, De Groot, and Hollander, 1990; Toussaint and Beck, 1992; Jang, Flynn, Costill, Kirwin, Houmard, Mitchell, and D'Acquisto, 1987). A swimmer can be tethered in a moving flume of water of known velocity. The applied force upon the tether can be measured, and C_d can be found by dividing the measured force by $\rho v 2S/2$. One convenient approach is to obtain the volumetric drag coefficient at a given velocity, which does not require measuring the surface area in contact with water. By assuming a slender cylinder for the shape of the swimmer, S is taken to be the 2/3 power of the volume of displaced water, $(m/\rho)^{2/3}$. Efficiency e less than one occurs because some of a swimmer's effort results in adding kinetic energy to water but not in providing useful forward thrust. The power ratio for two competitors in a hydrodynamic event, using the volumetric drag coefficient, for powers P_1 and P_2 is

$$\frac{P_1}{P_2} = \left(\frac{v_1}{v_2}\right)^3 \left(\frac{m_1}{m_2}\right)^{2/3} \frac{C_{d1}}{C_{d2}} \frac{e_2}{e_1}. \tag{3.8}$$

Elite female and male swimmers apply 100 to 300W of power to the surrounding water; but efficiency is low (Toussaint et al., 1990; Toussaint and Beck, 1992). Efficiency depends on the swimmers size, in much the same way that a large propeller is more efficient than a small one (Toussaint et al., 1991, 1990). A swimmer applies power using the factors listed in Table 3.6.

TABLE 3.6

Factors influencing the power output of a swimmer.

Physiology	
	Size and general fitness of the pool of athletes
	Lean body mass
	Muscle specific training
Techniques	
	Arm positioning and rotation
	Leg positioning and rotation
	Trunk positioning and rotation
	Breathing efficiently
	Turns
Coaching the technique	
	Record past performances
	Evaluate performances
	Train for proper technique
Quality of the venue and personal equipment	
	Swim suit drag
	Wave damping
	Water temperature
	Water clarity

3.4.2 MEASURING THE RATE OF IMPROVEMENT IN SWIMMING

Swimming is a timed event so that Equation 3.3 may be used to calculate $\%I/O$. Table 3.2 contains $\%I/O$ values for swimming. Men first swam in the 1896 Games; however, competition was in a Greek harbor at that time, with the current of the Seine for the Paris Olympics in 1900 and yard distances were used in 1904 rather than metric distances. Metric distances in pool competition started for men in 1908, the first data used for this study. Women entered swimming competition in 1912. The number of $\%I/O$ values used for Table 3.2 vary for men from 6 in 1912 to 16 in 2004 and for women from 2 in 1920 to 16 in 2004. The improvement for male swimmers followed the same pattern as for male running and jumping athletes during the WW1, WW2, and cold war periods. The negative improvement at Moscow in 1980 and the positive improvement at Los Angeles in 1984 are consistent with the dominance of Western-bloc male swimmers of that era who were absent at Moscow but present at Los Angeles. Female swimmers had a large $\%I/O$ value in 1920 in their second competition, as women played an increased role in domestic WW1 wartime efforts and entered swimming competition in much larger numbers than earlier. Women also showed WW2 wartime gains in 1948, compared to 1936 for most likely the same reason as after WW1, followed by consistently high improvement during the cold war. The positive improvement of 1980 compared to the negative improvement in 1984 is consistent with the dominance of Eastern-bloc women, present in 1980 and absent in 1984. Improvement rebounded in 1988 at the fully attended games.

Increased drug testing before and during the 1996 Games, coincided with negative improvement. A case in point is the criticism levied against Ireland's Michelle Smith in 1996, when she won gold medals in the 400m freestyle, the 200m individual medley, and the 400m individual medley, while passing drug tests. Smith had dramatically improved her personal best in each of those events, which caused some competitors to accuse her of masking use of performance-enhancing drugs. In fact, Smith's three winning times were rather mediocre: her three winning times would have won no gold in 1992, no gold in 1988 and only one gold in 1984. Her times won in 1996, partially from Smith's improvement and partially from the fact the typical women's times were 0.26% worse than in 1992.

Based on averages over each of the five periods, the rate of improvement in swimming has dropped consistently. Women improved more than men previously, but now improve at about the same rate. Average %I/O values since 1956 are about the same for swimming and jumping, both of which improved much more than for running. As with jumping, swimming is a very technical sport which affords more opportunity for improvement than running, which is dominated by foot-ground contact.

Table 3.3 contains the three best %I/O performances in swimming. These breakthrough performances were accomplished by one of several Hawaiian swimmers who dominated swimming in the early 1920s, Warren Kealoha, and two of a long line of colorful Australian swimmers, Andrew (Boy) Charlton and Dawn Fraser.

Earlier, we discussed England's Harold Abrahams, winner of the 100m run at the 1924 Chariots of Fire Olympics. Two famous swimming gold medalists at those same Games were the USA's Johnny Weissmuller, who won the 100m freestyle in 0:59.0, the first sub one minute swim for the 100m freestyle and Charlton, mentioned above, whose winning 1500m freestyle time was 20:06.6. Charlton left a 20 minute barrier as his legacy in winning. History has recorded the fall of the 20, 19, 18, 17, 16, and 15 minute barriers. Imagine trying to convince the 1924 mind, that a relay of 15 Johnny Weissmullers (15 Tarzans), each swimming 0:59 for 100m, would not win the 1500m individual swim in 2004, won in 14:43.40 or that the 100m freestyle would be won in 2004 with a time of 0:48.17. As done with Abrahams, the power equation allows us to give Weissmuller and Charlton their just due.

Let all body masses be 80kg to simplify the calculations. Let's start with the 2004 times. For 2004, suppose the volumetric drag coefficient is denoted $C_{d2} = 0.14$ while the efficiency is denoted $e_2 = 0.4$. The volumetric drag coefficient assumes that the whetted surface area is $(m/\rho)^{2/3}$ or 0.186m^2. An alternate method would be to find the actual surface area which probably would be much less while the drag coefficient would be much more. The product of those two terms would be about the same, but the volumetric method avoids finding the actual surface area. Using Equation 3.7, the 2004 100m freestyle winner produced 291W of power while the 2004 1500m freestyle winner produced 159W of power. For 1924, suppose the volumetric drag coefficient is denoted $C_{d1} = 0.154$, 10% higher than in 2004 due to the bulky drag-producing tank top shirt worn by male swimmers at the time. Also, let the 1924 efficiency be denoted $e_1 = 0.38$, 5% less efficient than in 2004 due to lack of wave damping lane markers and pool gutters. Further, pool clarity may have been worse causing the swimmer to follow less of a direct path down each lane. Weismuller would have produced 183W while Charlton would have produced 72.2W. What would those 1924 swimmers have done under 2004 conditions, assuming their output powers remained 183W and 72.2W respectively? In Equation 3.8, let the subscript "2" represent 2004 conditions while "1" represents 1924 conditions. With $P_1 = P_2$ and the masses being equal, the time t_2 under current conditions can easily be found in terms of the time t_1, under 1924 conditions, given that each velocity is distance/time. Then t_2 is simply $t_1 \left[(C_{d2}/C_1)(e_1/e_2) \right]^{1/3}$, which is $t_1 \times 0.95$. Thus, 15% of relative improvement in drag and efficiency translates into 5% improvement in time. Put another way, each 1% improvement in time requires a 3% improvement in drag and efficiency, for the same power output. Weissmuller's time becomes 0:56.05 which would have won as late as 1952 while Charlton's time becomes 19:05.7, which would have won as late as 1948. The significantly higher powers of the 2004 swimmers indicate the resulting increase in energy storage for today's swimmers, generated by increased pool time, increased numbers of repetitions, and increased muscle-specific weight work done in dry land training.

3.4.3 THE FUTURE OF SWIMMING

Swimming times generally improved in 2000 and 2004, while running and jumping performances generally improved only in 2004. Ostensibly, swimming is now a clean sport, given two Olympics of improvement in an era of intense drug testing. In fact, only one Olympic medal in swimming has ever been stripped (in 1972), compared to five medals in track and field and twelve medals in weightlifting. Swimming times should continue to improve, but at a lower rate than in the past, while women should improve at nearly the same rate as men. Since swimming, like jumping, is highly technical, the rate of improvement in swimming should continue to be about twice that of running and about that of jumping. Further, it is possible that the exponential growth in computer speed and storage capability will make it possible to identify improvements in stroke technique.

3.5 ROWING

3.5.1 THE PHYSICS OF ROWING

The force applied by rowers to propel a boat forward is opposed by the force of hydrodynamic drag as with swimming; therefore, Equation 3.7 and Equation 3.8 are applicable. For rowing, mass includes that of the rowers, boat, and oars. A computer simulation may be used (Tuck and Lazauskas, 1996) to estimate the drag coefficient of a racing shell, which is generally much less than for a swimmer. Efficiency e is much higher for rowing, compared to swimming, because in rowing only the oar surface generates kinetic energy lost to the surrounding water while a swimmer's entire body generates relatively more lost kinetic energy. Kinetic energy can be also lost in rowing by dragging an oar or being out of synchronization with team mates. Stefani and Stefani (2000) estimated the power applied per rower for all Olympic winning performances from 1908-1996, using drag coefficient values from Tuck and

Lazauskas (1996), athlete weights estimated from team rosters and boat weights provided by manufacturers.

For example, suppose the total mass of the boat, four male rowers, and the oars is 400kg, the volumetrically-calculated C_d is 0.028, efficiency is one, the course covers 2000m, and the volumetrically calculated whetted surface area S is $(400/1000)^{2/3}$ or 0.543m^2. In quad sculls each rower uses two oars. The winning quad sculls time in 1996 for men was 5:56.9. Using Equation 3.7, the total power output of the four rowers was 1338W or 334W per rower. For fours without a coxswain competition, with one oar per rower, the winning time for men was 6:06.4, giving a total power output of 1236W or 309W per rower. The greater power output in sculls competition indicates the relative advantage of each rower having a second oar.

Generally, quad sculls boats are faster than double sculls, which are faster than single sculls. In sweep-oars competition with one oar per rower, eights are faster than fours, which are faster than pairs. Equation 3.8 is helpful here. As mentioned earlier, an athlete's output power is generally in proportion to lean body mass. If we assume P_1/P_2 is approximately equal to the ratio m_1/m_2 for two boats with equal drag coefficients and equal efficiencies, the velocity ratio for those two boats, v_1/v_2, would be about equal to $(m_1/m_2)^{1/9}$. If m_1 is twice m_2, the velocity ratio should be about $2^{1/9}$ or 1.08. For example, in the 1996 Olympics, (winning women's quad sculls velocity)/(winning women's double sculls velocity) was $5.16/4.80 = 1.08$ while (winning women's double sculls velocity)/(winning women's single sculls velocity) was $4.80/4.42 = 1.09$, both agreeing with theory. The factors influencing the power output of a rower are listed in Table 3.7.

Rowing, like swimming, requires a variety of coordinated motions to effectively move through water.

3.5.2   MEASURING THE RATE OF IMPROVEMENT IN ROWING

Rowing is a timed event; therefore, Equation 3.3 may be used to calculate $\%I/O$ values. In order to find meaningful rates of improvement, it is necessary to have data for a consecutive set of performances, contested under reasonably equivalent conditions. Men first competed in rowing in 1900; however, the distances in 1900, 1904, 1908, and 1948 were not 2000m. Currently, all men's and women's events are contested over a 2000m course, usually on a lake. Although men's distances became 2000m starting with 1912 (except for 1948), the 1924 and 1936 values are not useful because the venues were rivers. In those cases, the velocity relative to the river bank that can be calculated from the official elapsed time is not the same as the velocity of water rushing alongside the rowing shell. Each power calculated using Equation 3.7 would have to be corrected for the current at the time of competition, data that is not available. Except for wind effects, which can be substantial, men's rowing conditions have been consistent since 1952. Women first competed in rowing in 1976; but the distance did not become 2000m until 1988. For this study, the number of $\%I/O$ values vary for men from seven in 1956 to eight in 2004 and for women from five in 1992 to six in 2004. The values are averaged per Olympics and for longer time periods in Table 3.8.

TABLE 3.7
Factors influencing the power output of a rower.

Physiology	
	Size and general fitness of the pool of athletes
	Lean body mass
	Muscle specific training, especially using a rowing ergometer
Techniques	
	Forward push/ recovery/ rotation of oar
	Coordination of hands, arm, legs, and back on the pull
	Stroke coordination with team mates
Coaching the technique	
	Record past performances
	Evaluate performances
	Train for proper technique
Quality of the venue and personal equipment	
	Rigging
	Flexible, lightweight clothing
	Low friction rolling seat
	Steer ability by the coxswain
	Low drag coefficient

Varying wind conditions render the $\%I/O$ value for a given Olympics somewhat imprecise; however, an average taken over several Olympics tends to smooth out wind effects. The average $\%I/O$ for men from 1956-1976 was 1.22%, higher than for running but not quite as high as for jumping and swimming performances over that cold war period.

The period from 1980-1988 requires special consideration. East German rowers dominated the 1980 Games, innovating with extensively more on- and off-water training than had been the norm previously. The average $\%I/O$ for 1980 was 7.42%, at a time when Western-bloc athletes were not present. We would expect most of that improvement to disappear when East Germany was absent at Los Angeles in 1984, but there was only a small negative improvement of -0.14%. How did the Western-bloc athletes respond? Starting with about 1976, the rowing ergometer was developed and then it became ubiquitous. To understand the change in the training mind set brought about by the rowing ergometer, consider an interview of Gordon Adam (Hodak, 1988), a gold medalist in rowing eights from the 1936 Games in Berlin. Adam recounts that, during the long boat trip to Europe, the cruise ship had an exercise room, containing a type of rowing machine that was popular at the time for casual exercise. The rowing coach forbade the use of that machine because he concluded that the machine was not calibrated properly, and he felt that the rowers would lose proper muscle coordination. The coach insisted that the only way to train for rowing was to row. The rowers exercised and ran for fitness while onboard. Adam

TABLE 3.8
Percent improvement per Olympiad for rowing.

Olympics	%I/O Men	%I/O Women	Olympics	%I/O Men	%I/O Women
1956	1.68		1980	7.42	
1960	9.92		1984	−0.14	
1964	−8.69		1988	1.46	
1968	3.97				
1972	4.21		1992	2.25	3.00
1976	−3.8		1996	−1.16	−1.60
			2000	0.89	1.14
			2004	−1.49	−0.39
1956-76	1.22		1980	7.42	
			1984-88	0.66	
			1992-04	0.21	0.52

and his eights teammates from the University of Washington crew, the Olympic trials winner, had time for on-water training before the Games, and they came away
with the gold medal. This anecdotal story reveals the mind set of competitive rowing
before 1980: the primary focus was on-water training, that was highly weather dependent, while running and cycling was a supplement. The rowing ergometer started
as a stationary bicycle with a friction mechanism and then advanced into a device
that provided rowing-muscle-specific exercise that was not weather dependent. That
device significantly increased effective rowing-specific-muscle training. Ergometers
have become so popular that there are national and international ergometer competitions. In Table 3.8, it is estimated that the rowing ergometer increased %I/O by
508%, based on the 1956-1972 average of 1.22% compared to the 1980 value of
7.42%.

The 1984-88 and 1992-2004 averages for men indicate reduced rates of improvement. For women, the 1992-2004 average of 0.52% is higher than for women in running, jumping, and swimming and slightly higher than for men. Table 3.3 contains
the best three %I/O event values in rowing, the highest two of which are from 1980
while the other was performed on Lake Albano, Italy in 1960, a year with favorable
rowing conditions.

3.5.3   THE FUTURE OF ROWING

Rowing shares with the other sports the trend that rates of improvement have generally declined over time, except, of course, for 1980. Another breakthrough as significant as the rowing ergometer is unlikely; since the racing shell already has very low
drag, and the ergometer already provides rowing-muscle-specific training. It is possible that the oars could be redesigned, but improvement there is not likely to exceed
5%. Improvement will probably continue near the 1992-2004 level in the short term,

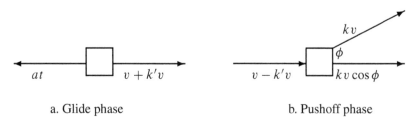

a. Glide phase   b. Pushoff phase

FIGURE 3.3
Velocities in speed skating looking downward.
Skater is denoted by the box. (a) Glide phase starts with velocity $v + k'v$, (a) Glide phase ends with velocity $v - k'v$, (b) Pushoff phase starts with velocity $v - k'v$, and (b) Pushoff phase ends with velocity $v + k'v$.

barring unusual wind conditions. The 1992-2004 average may somewhat underestimate a level of improvement that is sustainable for the next few years, since poor wind conditions negatively affected performances in 2004 at the Athens Olympics. Women and men are likely to improve at nearly the same rate.

3.6 SPEED SKATING

3.6.1 THE PHYSICS OF SPEED SKATING

A detailed analysis of the biomechanics of speed skating (Van Ingen Schenau, De Groot, and De Boer, 1985; De Boer, Cari, Vaes, Clarijs, Hollander, De Groot, and Van Ingen Schenau, 1987; De Boer, Ettema, Van Gorkum, De Groot, and Van Ingen Schenau, 1987; De Koning, De Groot, and Van Ingen Schenau, 1989, 1991) reveals that the athlete begins a race with an acceleration sprint phase after which it is not possible to push directly backward, due to the high forward velocity, typically more than 10m/s. The athlete then alternates between gliding and pushing off to maintain forward velocity. During the glide, the body is positioned to reduce aerodynamic drag. Following each glide, the athlete pushes off to alternating sides, each push off being at an angle from the forward direction. The joint rotations, translations, forces, and torques are similar to those employed in a vertical jump. The current clap skate allows use of strong extensor muscles generating more power than with the former rigid skate, where certain muscle movements had to be suppressed. An elite speed skater achieves increased velocity with the same amount of energy per stroke but with increased stroke frequency, so that power output is proportional to velocity.

Consistent with (Van Ingen Schenau et al., 1985; De Boer et al., 1987; De Koning et al., 1989, 1991), the physics of speed skating can be described using Figure 3.3, looking directly down at the speed skater whose *CG* moves from left to right. The glide phase is illustrated in Figure 3.3a, which starts at time $t = 0$, when the forward velocity of the *CG* is $v + k'v$. During the remainder of the glide phase, that forward velocity is reduced by the friction of the ice and by aerodynamic drag. Let at represent the combined effects of ice friction and drag, so that the forward velocity is given by $v + k'v - at$. Let the glide phase end at time T when aT equals $2k'v$. At that time, the forward velocity becomes $v + k'v - 2k'v$ or $v - k'v$. The average velocity during the glide phase is v, but at the end of the glide phase the skater needs to recover the lost velocity $2k'v$. A push off is illustrated by Figure 3.3b. Again looking down on the skater, initially moving at a forward velocity $v - k'v$, the left skate is lifted from the ice and the skater pushes off to the left. Relative to the skater's velocity at the end of the glide phase, the skater's *CG* moves with an additional velocity kv, oriented at an angle ϕ with respect to the forward direction. The forward component of that added velocity is $kv \cos \phi$. If we view the skater from the skater's side, as in Figure 3.3b, the horizontal and vertical velocities are as in Figure 3.1 and the skater's *CG* trajectory becomes the part of Figure 3.2 that spans L units, with v in those figures replaced by kv. The push off is completed when the left skate touches the ice. The skater attempts to cause $kv \cos \phi$ to be equal to $2k'v$. That choice makes the forward velocity at the end of the push off, which is the start of the next glide, to be $2k'v + v - k'v$ or $v + k'v$. The lost velocity has been restored. The skater again glides, pushes off to the right, again glides and pushes off to the left until the end of the straightaway. For the turn, where direction changes, a crossover stroke is used. Motion along the turn is beyond the scope of this discussion. Since we replace v with kv to replicate the work of the above authors, we can use Equation 3.2 with kv replacing v. The power output on the push off phase is given by

$$P = mgkv \tan \theta / 2e. \qquad (3.9)$$

A push off angle of about $50°$ is suggested in the references above. If $e = 1$, a forward velocity of $v = 10$m/s can be maintained by an 80kg speed skater with $\theta = 50°$ and $kv = 1.3$m/s on the push off, requiring an applied power during the push off of 607W. The skater rests somewhat during the glide. If the glide takes twice as long as the push off, the average power would be about 200W over each cycle, excluding power used to maintain the glide position. Based on the above discussion, the factors in Table 3.9 are important for proper speed skating performance. Speed skating is obviously a technically demanding sport with significant opportunity for improvement.

3.6.2   MEASURING THE RATE OF IMPROVEMENT IN SPEED SKATING

Speed skating is a timed event so that Equation 3.3 may be used to calculate $\%I/O$. Table 3.10 contains $\%I/O$ values for speed skating. Speed skating for men began with the first Winter Olympics in 1924 at Chamonix, France. Speed skating performances for men at Lake Placid in 1932 are not used, because the 1932 events employed a mass start as in current short track skating. Times were slowed by skating in a pack, rather than in a separate lane as in the other years in this study. Women entered speed skating competition at Squaw Valley in 1960. The number of $\%I/O$ values used for

TABLE 3.9
Factors influencing the power output of a skater.

Physiology	
	Size and general fitness of the pool of athletes
	Lean body mass
	Muscle specific training
Techniques	
	Sprint phase
	Push off angle, skate edges, body position, explosive extension
	Glide efficiency, body position
	Efficient crossover stroke on the turns
Coaching the technique	
	Record past performances
	Evaluate performances
	Train for proper technique
Quality of the venue and personal equipment	
	Skates
	Ice
	Flexible, warm, lightweight clothing

Table 3.10 vary for men from 4 in 1928 to 5 in 2006 and for women from 4 in 1964 to 5 in 2006. The best three performances in terms of %I/O are shown in Table 3.3.

WW2 appears to have negatively impacted men's speed skating in 1948, as happened for other sports discussed earlier. As with rowing, the nature of the venue can strongly affect performance. For speed skating, ice conditions have varied considerably. The venue has changed from outdoor lakes, to outdoor rinks to indoor rinks. The highest average %I/O for one Olympic competition was 5.20% for men at Cortina, Italy in 1956, where the venue was a frozen lake. Each morning a technician broke the first few feet of ice nearest the shore, resulting in a relatively hard and smooth ice surface, compared to earlier venues. Ice conditions at Calgary were also exceptional in 1988, resulting in the second highest improvement for men at 5.16% and the highest for women at 4.44%. Notice that the ice-related improvements in 1956 and in 1988 were not sustained in 1960 and 1992, respectively. Further, all three of the best %I/O values for speed skating in Table 3.2 happened under the excellent ice conditions at Cortina in 1956 and Calgary in 1988. By taking a long term average, these unusual environmental effects may be "averaged out" as was done for wind conditions in rowing, revealing purposeful improvement of the factors listed above.

3.6.3   PERIODS OF WINTER OLYMPIC HISTORY

The first two periods for Winter Olympic history may be chosen to be the same as the second and third periods of Summer Olympic history; that is, the period around WW2 covering four of the five Winter Games from 1928 -1952 (the 1932 results were

TABLE 3.10
Percent improvement per Olympiad for speed skating.

Olympics	%I/O Men	%I/O Women	Olympics	%I/O Men	%I/O Women
1928	-1.00		1980	3.79	4.22
1936	3.41		1984	-1.50	3.01
1948	-0.08		1988	5.16	4.44
1952	1.32		1992	-2.70	-4.02
1956	5.20		1994	3.86	2.84
1960	0.75		1998	2.49	3.27
1964	0.67	1.13	2002	3.14	3.17
1968	2.77	1.07	2006	-1.16	-2.30
1972	1.16	2.44			
1976	1.15	2.24			
1928-52	0.91		1980-92	1.19	1.91
1956-76	1.95	1.72	1994-06	2.08	1.75

not used as discussed above) and the period covering the six cold-war Games from 1956-1976. The next period for the Summer Games is 1980-1988, when the 1980 and 1984 Games involved boycotting while the 1988 Seoul Olympic competition was fully attended. The Winter Games did not suffer from boycotting in 1980 and 1984; hence, it is more useful to divide Winter Games from 1980-2006 into two sets of four. By averaging data for the four Games from 1980-1992, the effect of the unusually good ice conditions in 1988 can be combined with the effect of normal ice conditions in 1992. The recent period may be chosen to include the four Games from 1994 to 2006. Starting with 1994, the Winter Games began to be offered between the dates for Summer Games.

The %I/O values averaged over each period show more consistent improvement than for the other sports considered earlier. For men, the averages are 0.91%, 1.95%, 1.19%, and 2.08% respectively. For women, we have 1.72%, 1.91%, and 1.75%. The Games of 1998 and 2002 require special consideration. Prior to the 1998 Games in Nagano, Van Ingen Schenau was able to entice junior-level speed skaters in the Netherlands to use the so-called clap skate that had actually been invented around 1900, but the invention had remained unused. The clap skate has a hinge that allows the skating blade to remain in contact with the ice as the skater straightens the leg, thus employing strong extensor muscles; in much the same way as the Fosbury Flop allows a high jumper to use the same extensor muscles. The push-off angle from the horizontal is about 50° for proper use of both the clap skate and the Fosbury Flop. When the skate leaves the ice surface, a spring causes the skate to return to contact with the shoe, making the clapping sound which gives that skate its name. The success of junior-level skaters led to rapid acceptance and availability of that device for the 1998 Games at Nagano. All ten Olympic speed skating records fell, five for men and five for women. The effectiveness of the clap skate is estimated in Table 3.5. From 1956-1994 the average $\%I/O$ was 1.82%, averaging data for men and women. In 1998, the average $\%I/O$ for men and women was 2.88% (based on 2.49% for men and 3.27% for women), indicating an increase in $\%I/O$ of 58% when the clap skate was first introduced.

Improvement persisted at Salt Lake City in 2002 when all ten Olympic records again fell. In 2002, $\%I/O$ was 3.14% for men and 3.17% for women. The Salt Lake City Organizing Committee commissioned an indoor rink with state-of-the-art temperature and humidity control of rink air and temperature control of the rink ice. The excellent ice conditions and high altitude contributed to the high $\%I/O$ values that were not sustained at Turin, Italy in 2006. In fact, all ten winning performances in 2006 were worse than in 2002 due to less-than-ideal ice conditions.

3.6.4 THE FUTURE OF SPEED SKATING

As with rowing, in which wind effects must be considered, multi-Olympic speed skating averages must be taken to filter out the influences of unusually good ice conditions and unusually bad ice conditions so as to reveal actual improvement of the factors that influence speed skating performance. The long-term averages for speed skating do not exhibit the consistently reduced level of improvement that we have seen for running, jumping, swimming, and rowing events. The design of the clap skate hinge is still under development and further improvements are possible. Van Ingen Schenau et al. (1985); De Boer, Cari, Vaes, Clarijs, Hollander, De Groot, and Van Ingen Schenau (1987); De Boer, Ettema, Van Gorkum, De Groot, and Van Ingen Schenau (1987); De Koning et al. (1989, 1991) indicate a high level of understanding of speed skating bio-mechanics with a continued potential for improvement. The rate of improvement, appropriately averaged, should remain at more than one percent per Olympiad and women should continue to improve at about the same rate as men, in that opportunities for men and women have been, and continue to be reasonably equal in those countries winning the most speed skating medals.

3.7 A SUMMARY OF WHAT WE HAVE LEARNED

For those wanting to delve further into past performances in the Summer and Winter Games, Wallechinsky's two books (Wallechinsky, 2004, 2005) are recommended. We began our coverage of each sport by evaluating those physical laws that govern the motion of a runner, jumper, swimmer, rower, and speed skater. Those laws provide an understanding of the amount of power required to master each sport and thereby we obtained a list of factors that define success in each sport. Those same factors provide the opportunity for improvement in four areas: the physiology of each athlete, the techniques available to the athlete, coaching methods available to teach those techniques, and the quality of the competitive venue and equipment. The physics of running is dominated by the foot-ground reaction force, which delimits opportunity for improvement. Conversely, jumping, swimming, rowing, and speed skating are much more technically challenging and thereby offer more opportunity for improvement.

There have been four noteworthy technical breakthroughs. The rowing ergometer dramatically increased out-of-water rowing-muscle-specific training. The Fosbury Flop technique for the high jump and clap skates used in speed skating both enabled use of strong extensor muscles whose use had been suppressed by earlier methods. The fiberglass pole allowed more kinetic energy to be translated into potential energy enabling greater heights to be cleared in the pole vault.

Percent improvement per Olympiad, designated $\%I/O$ was used to evaluate the rate improvement. For all events in this study, the average value of $\%I/O$ was 0.70% for running, 1.56% for jumping, 1.54% for swimming, 1.25% for rowing, and 1.62% for speed skating. The events with more technical challenge, jumping, swimming, rowing, and speed skating, thus exhibited about twice the improvement of running. Athletes have often been held captive to outside influences beyond the control of sport. For example, there were the negative influences of WW1, WW2, and boy-cotts and the positive influence of the cold war period whereby Olympic success was considered to be a benchmark of Eastern-bloc versus Western-bloc dominance. Recently, certain event performances have become worse after significant efforts to punish doping were initiated. The attention of athletes was obviously gained in 1988 when Ben Johnson was disqualified for doping, after setting an apparent world record in the 100m run at Seoul. The aftermath of drug testing may be seen in that about two-thirds of the track and field winning performances in Seoul in 1988 would still have won at Sydney in 2000. This may well have occurred as more recent drug-free athletes have winning performances that are compared to those of earlier Olympians who had used performance-enhancing drugs.

Averages taken over several Olympics indicate that long-term improvement has been gradually declining in running, jumping, swimming, and rowing. Speed skating has had more consistent improvement.

Women had been improving faster than men in running, jumping, and swimming events and are now improving at about the same rate. I believe that women had less opportunity than men in those sports in the past and now have "caught up" in those three sports. In rowing and speed skating, women and men have improved at about the same rate. The countries which have dominated the medals in those sports have generally provided reasonably equal opportunity for both genders. In the future, women should improve at about the same rate as men for all of the sports covered in this study.

To close, let us return to 1891, five years before the start of the modern Olympic Games for perspective. In that year, a Dominican priest named Henri Didon, Prior of Arcueil College in France, spoke to his students at a sports club meeting. Didon taught that a person is not a success in life until that person has done the utmost to enhance the skills given to him while an athlete is not a true success until the athlete has tried the utmost to perform faster, higher, and stronger than that athlete's previous performances. Didon ended his speech with Citius, Altius, Fortius, the Latin translation of "faster, higher, stronger." To Didon, the athlete's antagonist is the athlete's own self. History records that Citius, Altius, Fortius was chosen as the Olympic Motto in an organizational meeting of the International Olympic Committee in 1894.

In every era we have already seen, proponents of each sport have followed Didon's admonition in that such proponents have worked step by step to improve the physiology of each athlete, to improve and create new techniques, to improve coaching methods and to improve the quality of the competitive venue and of athletic equipment. While the rate of improvement may be reduced in the future, Didon's challenge simply calls for improving performance, given the conditions in an athlete's own time. There is no reason to suggest that future athletes will be any less pro-active than their predecessors. There are likely to be technological breakthroughs, broken records, and perhaps another Beamonized performance. I believe that there are no limitations to the imagination and determination of an athlete to succeed, and that only the span of time beclouds our vision of achievements as memorable as those we have already seen.

REFERENCES

Croskey, M.A., P.M. Dawson, A.C. Luessen, I.E. Marohn, and H.E. Wright (1922). The height of the center of gravity of man. *American Journal of Physiology 61*, 171–185.

De Boer, R.W., J. Cari, W. Vaes, J.P. Clarijs, A.P. Hollander, G. De Groot, and G.J. Van Ingen Schenau (1987). Moments of force, power, and muscle coordination in speed skating. *International Journal of Sports Medicine 8*(6), 371–378.

De Boer, R.W., G.J. Ettema, H. Van Gorkum, G. De Groot, and G.J. Van Ingen Schenau (1987). Biomechanical aspects of push off techniques in speed skating the curves. *International Journal of Sports Biomechanics 3*, 69–79.

De Koning, J.J., G. De Groot, and G.J. Van Ingen Schenau (1989). Mechanical aspects of the sprint start in olympic speed skating. *International Journal of Sports Biomechanics 5*, 151–168.

De Koning, J.J., G. De Groot, and G.J. Van Ingen Schenau (1991). Coordination of leg muscles during speed skating. *Journal of Biomechanics 24*(2), 137–146.

Dubin, C.L. (1990). *Commission of Inquiry into the Use of Drugs and Banned Practices Intended to Increase Athletic Performance*. Ottawa: Canadian Government Publishing Center.

Hellebrandt, F.A. and E.B. Franssen (1943). Physiological study of the vertical standing of man. *Physiology Review 23*, 220–255.

Hodak, G.A. (1988). Gordon H. Adam, 1936 Olympic Games. Olympic Oral History Project, Amateur Athletic Foundation of Los Angeles.

Iwaoka, K., H. Hatta, Y. Atomi, and M. Miyashita (1988). Lactate, respiratory compensation thresholds, and distance running performance in runners of both sexes. *International Journal of Sports Medicine 9*(5), 306–309.

Jacoby, E. and B. Fraley (1995). *Complete Book of Jumps*. Champaign, IL: Human Kinetics.

Jang, K.T., M.G. Flynn, D.L. Costill, J.P. Kirwin, J.A. Houmard, J.B. Mitchell, and L.J. D'Acquisto (1987). Energy balance in competitive swimmers and runners. *Journal of Swimming Research 3*, 19–23.

Lerner, L. (1996). *Physics for scientists and engineers*. Boston: Jones and Bartlett.

Linthorne, N.P. (1999, 31 October - 5 November). Optimum throwing and jumping angles in athletics. In *Proceedings 5th IOC Congress on Sport Sciences with Annual Conference of Science and Medicine in Sport*, Sydney, Australia.

Lutoslawska, G., A. Klusiewics, D. Sitkowsi, and B. Krawczyk (1996). The effect of simulated 2 km laboratory rowing on blood lactate, plasma inorganic phosphate, and ammonia in male and female junior rowers. *Biology of Sport 13*(1), 31–38.

Maud, P.J. and B.B. Schultz (1986). Gender comparisons in anaerobic power and anaerobic capacity tests. *British Journal of Sports Medicine 20*(2), 51–54.

McArdle, W., F.I. Katch, and V.L. Katch (1981). *Exercise Physiology*. Philadelphia: Lea and Febiger.

McGinnis, P.M. (1991). Biomechanics and pole vaulting: Run fast and hold high. In *Proceedings of the American Society of Biomechanics, 15th Annual Meeting*, Tempe, AZ, pp. 16–17.

McGinnis, P.M. (1997). Mechanics of the pole vault take-off. *New Studies in Athletics 12*(1), 43–46.

Stefani, R.T. (1999). A taxonomy of sports rating systems. *IEEE Trans. On Systems, Man and Cybernetics, Part A 29*(1), 116–120.

Stefani, R.T. (2006). The relative power output and relative lean body mass of world and Olympic male and female champions with implications for gender equity. *Journal of Sports Sciences 24*(12), 1329–1339.

Stefani, R.T. and D. Stefani (2000). Power output of Olympic rowing champions. *Olympic Review XXVI-30*, 59–63.

Toussaint, H.M. and P.J. Beck (1992). Biomechanics of competitive front crawl swimming. *Sports Medicine 13*, 8–24.

Toussaint, H.M., G. De Groot, H.H. Savelberg, K. Vervoorn, A.P. Hollander, and G.J. Van Ingen Schenau (1988). Active drag related to velocity in male and female swimmers. *Journal of Biomechanics 21*, 435–438.

Toussaint, H.M., T.Y. Janssen, and M. Kluft (1991). Effect of propelling surface size on the mechanics and energetics of front crawl swimming. *Journal of Biomechanics 24*, 205–211.

Toussaint, H.M., W. Knops, G. De Groot, and A.P. Hollander (1990). The mechanical efficiency of front crawl swimming. *Medicine and Science in Sports and Exercise 22*, 402–408.

Tuck, L.O. and L. Lazauskas (1996, 30 Sept.- 2 Oct.). Low drag rowing shells. In *Third Conference on Mathematics and Computers in Sport*, Queensland, Australia, pp. 17–34. Bond University.

Van Ingen Schenau, G.J., G. De Groot, and R.W. De Boer (1985). The control of speed in elite female speed skaters. *Journal of Biomechanics 18*(2), 91–96.

Wakai, M. and N.P. Linthorne (2000, 7-12 September). Optimum takeoff angle in the standing long jump. In *Sports Medicine Book of Abstracts, 2000 Pre-Olympic Congress: International Congress on Sport Science, Sports Medicine, and Physical Education*, Brisbane, Australia.

Wallechinsky, D. (2004). *The Complete Book of the Summer Olympics, Athens 2004 Edition*. Sport Classic.

Wallechinsky, D. (2005). *The Complete Book of the Winter Olympics, Turin 2006 Edition*. Sport Classic.

Weyand, P.G., D.B. Sternlight, M.J. Bellizzi, and S. Wright (2000). Faster top running speeds are achieved with greater ground forces not rapid leg movements. *Journal of Applied Physiology 89*, 1991–1999.

4

Competitive balance in national European soccer competitions

Marco Haan

University of Groningen

Ruud H. Koning

University of Groningen

Arjen van Witteloostuijn

University of Groningen

ABSTRACT

According to popular belief, competitive balance in national soccer competitions in Europe has decreased due to the Bosman ruling and the introduction of the Champions League. We test this hypothesis using data from seven national competitions, for a host of indicators. We find some evidence for competitive balance having decreased in England, and weak evidence for it having decreased in Netherlands and Belgium. For Germany, France, Italy, and Spain, we find no consistent change whatsoever. We use factor analysis to examine whether our measures of competitive balance can be condensed in a limited number of factors.

4.1 INTRODUCTION

A world without soccer is unimaginable. The world soccer association FIFA has more members (207) than the United Nations (192). In most countries, the national soccer association is the largest sports association, soccer is the most broadcasted sport on television, and soccer players are the best paid athletes. With the 2002 World Cup having been organized in Asia, and the 2010 World Cup being played in Africa, it is likely that global interest in soccer will continue to increase.

Simultaneously with the growth of interest in soccer, the sport has developed itself as a professional activity. Professionalization and commercialization of soccer have taken off later than in American professional sports such as baseball, basketball, American football, and ice hockey (Szymanski and Zimbalist, 2003). Nowadays, soccer is a global industry despite its roots as a recreational activity. The increase of the economic value of soccer has changed both the game itself, and the organization of games and leagues. For example, the European Commission takes an active involvement in determining the system of transfer fees to be paid when a player leaves a team despite an ongoing contract. In The Netherlands, the bid book for the sale of television broadcasting rights was examined by the anti-trust authority before the actual procedure to sell the rights started. This increased attention for soccer by regulatory agencies is often justified by pointing to the need to maintain some level playing field between teams. This is one specific feature of sports industries in general and soccer leagues in particular: competition is their very product. Sports leagues need to produce "competitive excitement" to survive. Without competitive excitement, a sports league would be dull: after all, then there is not much that is attractive to customers (i.e., fans). This is in contrast with nonsports industries. The sports industries' uniqueness in the business world as a producer and seller of competition makes them particularly interesting from an antitrust perspective. A complication in the case of soccer is that there are at least two relevant competitions: national leagues such as The Premiership in England, the Serie A in Italy, and the Primera Division in Spain, but ever larger economic and sportive significance is attached to the main international league: the Champions League organized by the European soccer association UEFA. There is also a third level of competition, the competition between national teams. In this chapter we focus only on competitive balance in national leagues.

Especially at a national level, fans, media, and policy makers often worry about a decrease in competitive balance. Also, smaller teams like to stress the importance of competitive balance when new rules concerning the division of the proceeds of broadcasting rights are determined. However, there seems to be relatively little research into the development of competitive balance over time, and the consequences of two major shocks to competitive balance: introduction of the Champions League in 1992, and the Bosman ruling of 1995. The introduction of the Champions League has given teams the opportunity to earn relatively large sums of money in that competition. As a consequence, national competitive balance could be disrupted. The Bosman ruling has made players free agents upon expiration of their contract. As a result of this, smaller teams in particular, have lost income from transfer fees as an important source of revenue. The theoretical consequences of these shocks are examined in Haan, Koning, and van Witteloostuijn (2007).

In this chapter, we study the development of national competitive balance over time, for seven different countries: Belgium, England, France, Germany, Italy, The Netherlands, and Spain. Also, we examine the development of competitive balance at an international level by looking at European competitions for club teams. Because there is no general agreement in the literature concerning the measurement of competitive balance in soccer, we introduce and discuss different measures of competitive balance, and we try to condense these measures as well.

4.2   MEASUREMENT OF COMPETITIVE BALANCE

Most studies of competitive balance in soccer take a somewhat different view of competitive balance than we do here. Usually, competitive balance is not measured at the level of the league, but as the uncertainty of outcome of a game which is then often used to explain attendance. Examples of such studies are Chapter 10 and

Chapter 11 in this book, as well as Borland and Macdonald (2003) and Szymanski (2003). Studies that analyze competitive balance in soccer on a league level are for example Dobson and Goddard (2001) and Koning (2000). In both cases, a model is used to estimate quality parameters that characterize teams, and these estimated quality parameters are used to assess changes in competitive balance. In their study, Dobson and Goddard conclude that the importance of home advantage is decreasing over time in English soccer, and that results are characterized by negative short run persistence effects. Koning (2000) concludes that in the long run no systematic differences in team quality can be identified for The Netherlands.

In the context of American baseball, Vrooman (1996) distinguishes between three dimensions of competitive balance:

- closeness of the league within a given season;

- dominance of large market teams;

- continuity of performance.

The first two dimensions are within-season concepts of competitive balance, the third dimension captures a more dynamic aspect of competitive balance. The measures in Dobson and Goddard (2001) and Koning (2000) capture the first dimension only. Even though Vrooman operationalizes these concepts in baseball, they are equally applicable to soccer. In the remainder of this section we discuss measures that capture these three dimensions.

To measure competitive balance, it is not sufficient to calculate winning percentages, as is done by Quirk and Fort (1992) for American sports. First of all, approximately 25% of all soccer games end in a draw, and second, home advantage is an important characteristic of soccer (see also Chapter 12). We prefer to separate variation in home advantage from variation in team quality, because they are driven by different structural factors. Different models have been proposed to estimate the quality of teams, some focus on scores directly (Maher, 1982), others focus on the final outcome of the game (home, draw, or loss, see for example Clarke and Norman, 1995). Here, we use the Clarke-Norman model, because of its simplicity, and the easy interpretation of the parameters.

The Clarke-Norman model is specified as follows. Consider a game between teams i (home) and j (away). The goal difference of the final outcome of the game is denoted by GD_{ij}, which is positive if the home team wins, zero in the case of a draw, and negative if the away team wins. Goal difference is modelled as

$$GD_{ij} = h_i + \theta_i - \theta_j + \epsilon_{ij}, \tag{4.1}$$

where h_i is the home advantage of team i, θ_i the quality of team i, and θ_j is the quality of the away team j. All other factors determining the result (e.g., injuries, fatigue due to a game a few days earlier, etc.) are captured in the error term ϵ_{ij}, which is assumed to follow a normal distribution: $\epsilon_{ij} \sim \mathcal{N}(0, \sigma^2)$. Home advantage h_i is the margin by which the home team is expected to win, if it were to play an opponent of equal quality: $E(GD_{ij}) = h_i$ if $\theta_i = \theta_j$. The quality parameters θ_i and

θ_j are not identified in model (4.1): both parameters can be increased by a constant c, without affecting the probability distribution of the observable outcome variable GD_{ij}. We normalize the θs by imposing $\sum_{i=1}^{n} \theta_i = 0$, with n the number of teams. This means that average quality is zero, so the θ-parameters measure quality with respect to a hypothetical team with quality zero. A team with a positive θ is better than average, a team with a negative θ is a below average team. θ_i itself can now be interpreted as the expected win margin if team i would play an average team ($\theta_j = 0$) on a neutral venue ($h_i = 0$). Model (4.1) can be estimated by least squares. Even though the dependent variable is discrete, the model provides a reasonable fit to the data.

Competitive balance in a league can now be measured as the variation of the quality parameters θ. Of course, one can argue that the quality of soccer teams is better measured by their wage bill. We are not able to pursue this idea any further due to lack of information on financial data of teams. On the other hand, variation of the quality parameters θ provides an intuitively appealing measure of competitive balance: if all teams were equally good, all θs would be zero, and hence there would be no variation in the quality parameters. Results would be driven by variation in home advantage (h_i) and chance (ϵ_{ij}) only. In the next section we estimate variation of the quality parameters by the standard deviation of the estimated θs.

One could argue as well that differences in home advantage is a component of competitive balance that should be taken into account, see Forrest, Beaumont, Goddard, and Simmons (2005). With the Clarke-Norman model, this is straightforwardly done by looking at the variation of the home advantage parameters h_i. Note that, on a theoretical level, instruments to influence competitive balance due to home advantage will be different from those to influence the quality of teams.

A measure of the second dimension of competitive balance (dominance of large market teams) is the concentration ratio. This ratio is calculated as the number of points of the best k teams divided by the maximum number of points these teams could have obtained during the season. This is

$$CR_k = \frac{\sum_{i=1}^{k} P_{(i)}}{Wk(2n - k - 1)}, \tag{4.2}$$

with $P_{(i)}$ the number of points obtained by the ith best team, W the number of points per game won (which is 3 in modern soccer), and n the number of teams in the competition. The ratio is 1 for the k best teams if each of these teams win all games against teams that are lower in the ranking. Roughly speaking, CR_k is the number of points that the k best teams have obtained, as a fraction of the maximum number that they could have obtained. CR_k is an interesting measure of balance in a competition because it is believed that recently the top teams have improved so much over the lesser teams (for example, because of Champions' League revenues and because of better sponsor contracts) that they form a league of their own.

Finally, we propose to measure the last aspect of competitive balance (continuity of performance) by looking at the year-to-year ranking in a league. Year-to-year variation in ranking can be measured by calculating the number of changes in the ranking. Let r_{it} be the ranking of team i in year t, so that $r_{it} = 1$ for the team winning the competition, etc. A measure of change between two seasons is then

$$DN_t = \sum_i |r_{it} - r_{it-1}|. \tag{4.3}$$

Note that this measure treats moves up and down identically. Of course, DN_t depends on the number of teams in a competition. If a league consists of two teams, DN_t is either 0 or 2, in a league of three teams DN_t is either 0, 2, or 4, etc. If a league has an even number of teams, the maximum of DN_t is $\frac{1}{2}n^2$, with n the number of teams. A normalized measure of dynamics in a league is now

$$DN_t^* = \frac{2}{n^2} \sum_i |r_{kt} - r_{kt-1}|. \tag{4.4}$$

An alternative to this approach is to measure year-to-year (rank) correlation coefficients to measure consistency of ranking over time. DN_t^* assumes that rankings can be interpreted on an interval scale. In case one is reluctant to make such an assumption, variation of ranking over time can also be measured by calculating Spearman's rank correlation coefficient.

4.3 EMPIRICAL RESULTS

The development of the variance of the estimated θs over time for our seven different countries is graphically depicted in Figure 4.1. If quality differences between teams in a national competition would have increased over time, the lines in Figure 4.1 should be sloping upward. In Figure 4.1 and later figures we aid the interpretation by drawing a line which is a local regression of the variable graphed on time (see for example Hastie, Tibshirani, and Friedman, 2001).

The development of the coefficient of determination of model (4.1) over time is available on the website www.statistical-thinking-in-sports.com. From the standard F-test on significance of a regression for each country, we find that R^2 deviates significantly from 0 for all countries for all seasons. Hence, the parameters θ_i and h_i jointly differ significantly from 0. In Germany and Spain the competition has become less predictable over time as judged by the decrease of R^2 over time in these countries. Still, typical values of R^2 in recent seasons for these countries are in the $0.28 - 0.32$ range. The average value of R^2 for The Netherlands is 0.40. So 40% of the variation in goal difference in the Dutch soccer league can be explained by a model as simple as (4.1).

It appears from Figure 4.2 that the concentration ratio (4.4) has not changed much over time in the countries we examine. The variation of CR_4 from year to year is of much larger magnitude than any recent trend that can be discerned. But the concentration ratio pertains to the strength of the top four teams; it does not look at the identity of those teams. Do the same teams compete for the championship each year, or do the identities of the best teams change? This issue is addressed in Table 4.1, where we give (for each country) the number of teams that have made up the top

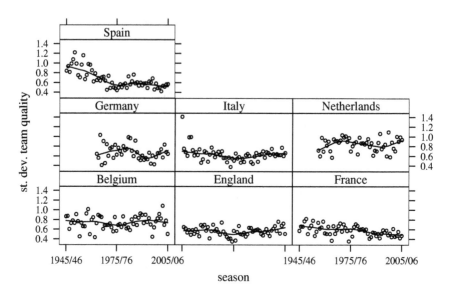

FIGURE 4.1
Variation of quality over time.

four in the end-of-season ranking, and the four teams that have appeared in the top four of the ranking most often. So for The Netherlands we see that 24 different teams have had a ranking in the top four during any season between 1955/56 and 2005/06. The teams with most appearances in the top four are: Ajax, Feyenoord, PSV, and Twente. Ajax has ended the competition in the top four 90% of all seasons, Feyenoord in 84%, etc. In Table 4.1 we see noticeable differences between countries. In The Netherlands there are three teams that make up the top, in Spain two teams, and in Belgium and Germany there is one team that has been consistently at the top. In France, though, the team that ended in the top four most often (Monaco), finished the season more frequently *outside* of the top four than in the top four. Moreover, it should be noted that in Belgium and England there are many more teams that have ended in the top ranking at some moment in time than in Germany, Italy, and The Netherlands. Apparently, the pool of potential national top teams is larger in Belgium and England than in the other countries. We conclude from Table 4.1 that the top is most concentrated in The Netherlands, followed by Italy and Spain. However, from Figure 4.2 it is clear that the quality difference of these teams with those that are ranked lower has not increased over time. The concentration at the end of the nineties is not different from the concentration during the early seventies.

In Figure 4.3 we graph the development of Spearman's correlation coefficient and the development of the year-to-year change of ranking DN_t^* in the seven leagues we analyze in this paper. From the smooth curves in the plots we see that there are no long-term trends: it is not the case that fewer teams change places at the end of the

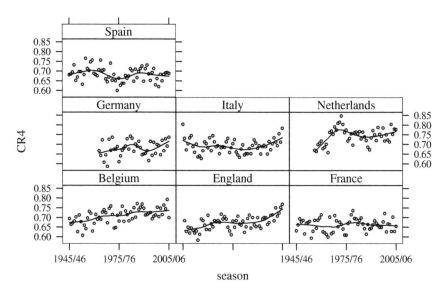

FIGURE 4.2
Concentration ratio CR_4 over time.

TABLE 4.1
Teams with most rankings in the top four (1945/46-2005/2006).

Country	Number of teams	Best four teams
Belgium	32	Anderlecht (93.4%), Standard Luik (56.7%) Club Brugge (555%), FC Antwerp (23.3%)
England	34	Manchester United (58.3%), Liverpool (55%) Arsenal (43.3%), Tottenham Hotspur (26.7%)
France	28	Monaco (43.3%), Bordeaux (41.7%) Olympique Marseille (33.3%), Nantes (30%)
Germany	21	Bayern München (79.1%) Borussia Dortmund (37.2%) Borussia Mönchengladbach (34.9%) Werder Bremen (34.9%)
Italy	20	Juventus (80%), AC Milan (71.7%) Internazionale (68.3%), Fiorentina (36.7%)
Netherlands	24	Ajax (90%), Feyenoord (84%) PSV (80%), Twente (24%)
Spain	20	FC Barcelona (88.3%), Real Madrid (85%) Atletico Madrid (51.7%), Valencia (41.7%)

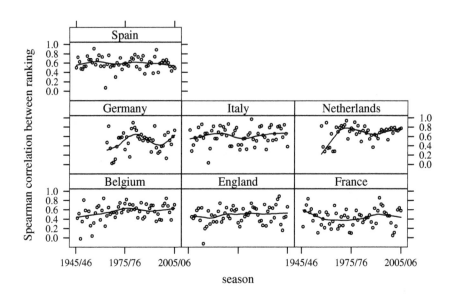

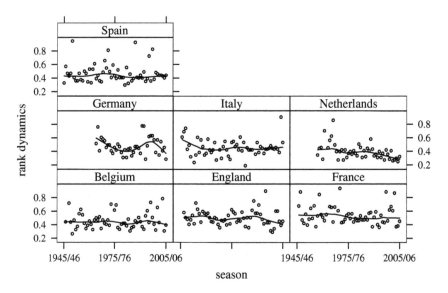

FIGURE 4.3
Dynamics of position, measured by Spearman's correlation coefficient (top) and
DN^* (bottom).

TABLE 4.2
p-values of Mann-Whitney test for no change.

	σ_θ	σ_h	CR_4	S	DN^*
Belgium	0.938	0.010	0.854	0.737	0.380
England	0.121	0.097	0.002	0.660	0.029
France	0.150	0.286	0.370	0.938	0.752
Germany	0.286	0.776	0.413	0.304	0.303
Italy	0.020	0.391	0.023	0.329	0.504
Netherlands	0.452	0.623	0.246	0.114	0.006
Spain	0.068	0.698	0.654	0.317	0.369

sample period than at the beginning of the sample period. However, there is clearly year-to-year variation. In some seasons teams change only a few places; in other seasons they change more places in the final ranking.

To test statistically whether competitive balance has changed in each country, one would ideally like to test two hypotheses: did competitive balance change after the introduction of the Champions League in 1992; and did it change after the Bosman ruling in December 1995? However, the period after the introduction of the Champions League and before the Bosman ruling is short, three seasons only. Of course, one could make strong assumptions about the effect of the introduction of the Champions League (for example, the effect is of equal magnitude in all countries), but we are reluctant to do so. The Spanish Primera Division has very different characteristics from the Belgian Eerste Klasse. Both events, though, are assumed to have had a similar effect on competitive balance: most people would argue that competitive balance has deteriorated over time. In any case, we can test the hypothesis whether or not balance before 1992 differs significantly from competitive balance after 1995. We do this by using a Mann-Whitney test, for each of the five measures of competitive balance ($\sigma_\theta = \sqrt{\text{var}(\theta_i)}$, $\sigma_h = \sqrt{\text{var}(h_i)}$, CR_4, S, and DN^*), and for each country.

In Table 4.2 we give the *p*-values corresponding to the hypothesis of no change. Most of the results in the table indicate no change of competitive balance, but there are a few exceptions. Variation in home advantage has increased in Belgium, and top teams have become more dominant in England. Also DN^* has decreased significantly in England. The results for Italy are not consistent with each other: variation of team quality σ_θ has increased, suggesting an increase of competitive balance. On the other hand, top teams have become more dominant, as judged from the increase of CR_4. Finally, dynamics of ranking has decreased significantly in The Netherlands as well.

We thus find that for most countries, most of our measures do not indicate a change in competitive balance due to the Bosman arrest and the introduction of the Champions League. For England we do find some evidence for a decrease in competitive balance: two out of five measures are significant and have the right sign. For the Netherlands and Belgium there is only weak evidence for a decrease, with one mea-

TABLE 4.3

Factor loadings, England and Belgium.

	σ_θ	σ_h	CR_4	R^2	σ_θ^{OP}	S	DN^*
England, two factor model							
factor 1	0.637	−0.028	0.934	0.688	0.992	0.061	−0.127
factor 2	0.166	−0.028	0.074	0.168	0.058	0.612	−0.989
England, three factor model							
factor 1	0.664	0.010	0.940	0.689	0.994	0.066	−0.140
factor 2	−0.135	0.065	−0.057	−0.162	−0.049	−0.616	0.987
factor 3	0.414	0.687	0.085	−0.039	−0.073	−0.082	−0.045
Belgium, one factor model							
factor 1	0.779	0.031	0.801	0.653	0.976	−0.042	−0.050
Belgium, three factor model							
factor 1	0.813	0.064	0.781	0.656	0.992	−0.009	−0.067
factor 2	0.239	0.995	−0.005	0.042	−0.067	0.132	0.138
factor 3	0.340	−0.029	−0.158	0.267	−0.086	0.636	−0.485

sure being significant. For France, Germany, Italy, and Spain, the evidence is either nonexistent, or contradictory.

4.4 CAN NATIONAL COMPETITIVE BALANCE MEASURES BE CONDENSED?

In the previous section, we used several measures of competitive balance. It would be interesting to see whether these indicators measure essentially the same variable, or whether there really are fundamentally different dimensions to competitive balance, as e.g., Vrooman argues. For that reason, we do a factor analysis in this section.

In fact, we can still increase the number of competitive balance indicators. For example, to what extent does luck determine results? This is measured by the coefficient of determination of regression (4.1). Also, we can measure quality in a slightly different way, using an ordered probit model that is not based on goal difference, but on the outcome of the game (home win/draw/home loss). Details of this approach are in Koning (2000) and Chapter 6 in this volume. The standard deviation of the quality parameters of this ordered probit model is denoted by $\sigma_\theta^{OP} = \sqrt{\text{var}\theta_i^{OP}}$. So by now, we have seven indicators: quality variation based on model (4.1), variation of home advantage, CR_4, Spearman's rank correlation S, dynamics of ranking DN^*, quality variation based on the ordered probit model, and the R^2 of regression (4.1).

Using seasons since 1976-1977 (as in the previous section), we examined whether the observed measures can be described by a limited number of factors. For each country, we estimated a factor model, with one, two, and three factors. The factor model was estimated by maximum likelihood, and interpretation of the results were simplified by the varimax rotation (which tries to estimate factor loadings such that a few loadings are as large as possible, and the others are small).

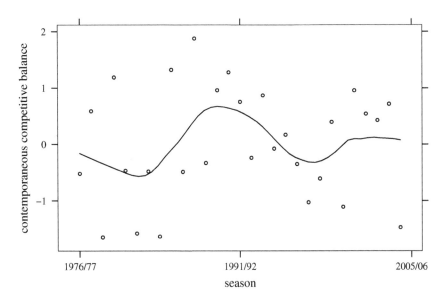

FIGURE 4.4

Factor score over time for the one factor model for Belgium.

Selected results are presented in Table 4.3. First, we look at the loadings for England. A two factor model is sufficient to model the correlation structure of the seven indicators, as is judged by the p-value of the test that examines sufficiency of the number of factors included, 0.234. The first factor can be interpreted as contemporaneous competitive balance. The second dimension of Vrooman (dominance of large market teams) is not identified as a separate factor, but instead it has a high loading on the first factor, just as σ_θ^{OP}. The second factor measures persistence of performance, Vrooman's third dimension. If we estimate a three factor model, one additional factor is identified: home advantage. For most countries, we find that home advantage appears as a separate factor in three factor models.

The second set of results are for a small league: the Eerste Klasse in Belgium. Only one factor is necessary to describe the correlations, and again this factor can be interpreted as contemporaneous competitive balance. Upon estimating a three factor model, we find again that the loadings can be interpreted as contemporaneous competitive balance, home advantage, and persistence of performance. Results for other countries are available on www.statistical-thinking-in-sports.com, and are roughly similar.

In Figure 4.4 we show the development of factor scores over time of one factor model for Belgium. In a way, this single graph summarizes seven graphs of competitive balance indicators. Unsurprisingly, no systematic trend over time is apparent.

4.5   CONCLUSION

In this chapter we have looked at the development of competitive balance over time. According to popular belief, competitive balance has changed noticeably due to two major changes in European soccer: introduction of the Champions League in 1992, and the Bosman ruling of 1995. We have discussed a number of empirical constructs to measure three dimensions of competitive balance: closeness of the league within a given season, dominance of large market teams, and continuity of performance. No across-the-board deterioration of competitive balance in seven national European leagues has been found, although there have been a few statistically significant changes. In the final section we examined whether all the different measures of competitive balance can be condensed in a limited number of factors. The results obtained from factor analysis with a varimax rotation are promising. It is especially noteworthy that the dimension "dominance of large market teams" does not appear as a separate factor, and that home advantage does.

REFERENCES

Borland, J. and R. Macdonald (2003). Demand for sport. *Oxford Review of Economic Policy 19*, 478–502.

Clarke, S.R. and J.M. Norman (1995). Home ground advantage of individual clubs in English soccer. *The Statistician 44*(4), 509–521.

Dobson, S. and J. Goddard (2001). *The Economics of Football*. Cambridge: Cambridge University Press.

Forrest, D., J. Beaumont, J. Goddard, and R. Simmons (2005). Home advantage and the debate about competitive balance in professional sports leagues. *Journal of Sports Sciences 23*(4), 439–445.

Haan, M.A., R.H. Koning, and A. van Witteloostuijn (2007). The effects of institutional change in European soccer. Mimeo, University of Groningen, The Netherlands.

Hastie, T., R. Tibshirani, and J. Friedman (2001). *The Elements of Statistical Learning*. New York: Springer.

Koning, R.H. (2000). Balance in competition in Dutch soccer. *The Statistician 49*(3), 419–431.

Maher, M.J. (1982). Modelling association football scores. *Statistica Neerlandica 36*, 109–118.

Quirk, J. and R.D. Fort (1992). *Pay Dirt: The Business of Professional Team Sports*. Princeton, NJ: Princeton University Press.

Szymanski, S. (2003). The economic design of sporting contests. *Journal of Economic Literature 41*, 1137–1187.

Szymanski, S. and A. Zimbalist (2003). *National Pastime*. Washington DC: Brookings Institution Press.

Vrooman, J. (1996). The baseball's player market reconsidered. *Southern Economic Journal 63*(2), 339–360.

5

Statistical analysis of the effectiveness of the FIFA World Rankings

Ian McHale
University of Salford

Stephen Davies
University of Salford

ABSTRACT

It is impractical to operate a world soccer league of national teams, given that there are 205 of them. Fans' curiosity regarding the relative strength of the team they support can therefore only be satisfied by consulting the FIFA World Rankings. But how reliable are these rankings as a guide to relative strength? An obvious criterion for helping to answer this question is their ability to forecast outcomes when two teams actually play each other. In this chapter we build a forecasting model for the outcome of soccer matches between national teams, and assess the extent to which the information it includes has been given appropriate weight in the FIFA rankings.

5.1   INTRODUCTION

In August 1993, the Fédération Internationale de Football Association (FIFA), the governing body of the world's biggest and most popular sport, soccer, published the first world ranking list of national soccer teams. Since its inauguration the rankings list has been inspected and scrutinized by fans, pundits, team coaches, players, and governing bodies across the globe. The rankings have come under much criticism from interested parties, mainly due to anecdotal anomalies in the ordering/positioning of some teams, but there also has been a debate on how ranking points are awarded to teams. The condemnation of the rankings meant that in 1998 FIFA decided to make "minor alterations" to their system as the original rankings process was "in need of improvement." Again, in 2006 following the World Cup Finals held in Germany, FIFA decided the time had come to once again alter the rankings system and the FIFA World Rankings underwent a comprehensive revision. In its relatively short existence, the FIFA world rankings list has therefore been the subject of much criticism. But was it justified? Were the changes FIFA made in response to such criticism necessary?

Although no formal statement for the objective of the rankings exercise was made, FIFA does state that the rankings provide "comparisons of the relative strengths of internationally active teams." It seems an intuitive step, therefore, that the rankings could be used to forecast match outcomes. This, of course may not be the sole objective of FIFA's rankings, since there may be some other, less transparent, purposes for them. For example, interest in soccer in a country relatively new to the game may be generated if the national team were given an inflated ranking.

The FIFA World Rankings list has provided seemingly illogical placement of some countries. A well publicized example is that of the rating of the USA. Prior to the 2006 World Cup, a rankings list showed the USA, a traditionally (relatively) weak team, ranked seventh in the list, and above soccer giants England, France, Italy, and Germany. The USA has never performed well in a major tournament and had recently been beaten comprehensively at home by England. One such result by itself does not necessarily indicate a problem since there is considerable noise in outcomes of football matches. But how do we know that such results represent genuine statistical variation and not some underlying flaw in the rankings system? No rankings system could provide perfect predictions. After all, the uncertainty in match outcome is a key contributing factor to the popularity of soccer. However, one would presume that the rankings system would provide good forecasts. The USA once again failed to progress to the latter stages of the World Cup held in 2006 despite its high world ranking of seventh.

If FIFA's rankings processed and weighted appropriately all relevant information then a forecasting model based on the rankings could not be improved by adding more information in the form of other variables. Past results for instance should be fully accounted for in the rankings, and one may infer that adding past results to the model would not improve forecasts. During the period 1998 to 2006, the rankings system was unchanged, and it is this period we focus on throughout this chapter. We attempt to answer the questions: how good are the FIFA World Rankings at forecasting match outcomes; and what information is the FIFA World Rankings not processing?

The rest of this chapter is structured as follows. First we introduce the current process used to produce the rankings and note implications of the FIFA World Rankings. Next we introduce the data set we have used in the modelling and some preliminary analysis. We then present our forecasting model.

5.2   FIFA'S RANKING PROCEDURE

For brevity, we provide a summary of the process used to produce the rankings during the 1998 to 2006 period. Full details of the ranking process can be found on the FIFA website[*]. The key points are:

- Points are awarded dependent on result and opposition strength. More points are awarded for a win than a draw, and more for a draw than a loss. In addition,

---

[*]www.fifa.com/en/mens/statistics/rank/procedures/0,2540,3,00.html.

more points are awarded for a result against a strong team than a weaker team;

- Teams are awarded more points as they score more goals, and similarly are awarded points for not conceding goals. The first goal scored is worth more than any subsequent goals scored. To reward and encourage attacking play, goals for are awarded more points than goals against;

- The away team receives bonus points, to account for any home/away bias;

- Weightings are applied to the points awarded depending on the type of match. Competitive matches in finals of tournaments are awarded the highest weighting, decreasing through qualifiers and friendlies;

- Weightings are also applied to the points awarded based on regional strength factors. Of the six confederations, matches between teams in UEFA (the European confederation) are weighted highest, followed by matches between teams from CONMEBOL (the South American confederation). If a match consists of teams from different confederations, then the average of the two weightings is used;

- Each of the previous eight years' match results are then weighted, with the most recent years results given greatest weighting, decreasing to the eighth year.

The process used by FIFA seems to have several subjective elements and no justification is given for the process. The weightings used on both the regional strength factors and past results do not seem to be based on objective elements. Similarly, bonus points for the away team and the actual points awarded for a win/draw/loss seem not to have any quantitative basis, or if one is used it is not highlighted by FIFA. Further, the decreasing weights applied to results from past years do not seem to have been based on any empirical work.

The regional strength weightings are an attempt to rectify a bias present in teams' opponents. Teams from the same confederation play each other more frequently than they play teams from other confederations due to the way groups are organized for tournament qualifying matches and the convenience of travelling shorter distances for friendly matches. Generally, the European teams are much stronger than those from the other confederations, and thus the European teams are, on average, playing stronger teams. The implication for an average European team is that their match results will look poor unless close account is taken of the average strength of their opposition.

5.3 IMPLICATIONS OF THE FIFA WORLD RANKINGS

The FIFA world rankings can have considerable effects. For example, the UK government uses the rankings to assess a soccer player's eligibility to be granted a work permit. One high profile example of a case in which a questionable consequence occurred was that of Liverpool Football Club's Chilean winger, Mark Gonzalez. Gonzalez had played for his national team on many occasions, and, although he was widely regarded as Chile's best young player, was not granted a work permit and could not play for Liverpool. The decision was based on the fact that his national side had, on average, been ranked outside the top 70 national teams. In fact Chile had an average ranking of 74th over the two years previous to the decision being made. In contrast to this, in 2004 Marton Fulop, a Hungarian goalkeeper was awarded a work permit to play for Tottenham Hotspur, based on the fact that Hungary had a ranking inside the top 70 teams. In fact Hungary was ranked 69th at the time and despite playing for an apparently high ranking country, Fulop has not yet played for the Tottenham first team and has spent much of his time in England on loan at lower division clubs. By basing such decisions on the FIFA World Rankings, the UK government is assuming the rankings are as accurate as they could be. These decisions affect individual careers, and in the interest of fairness, any tools used in the decision making process should be as accurate as possible.

5.4   THE DATA

We have collected a total of 8782 international soccer results from two main sources. The data for the period 1993-2001 were obtained from the archive of International Soccer Results[†] and the data for the period 2001-2004 were obtained from the RSSSF archive[‡]. Data on the FIFA world rankings were collected from the FIFA website[§] for each month during the years 1993-2004.

In addition to the results and rankings data, we used geographical coordinates to calculate approximate distances travelled by teams and/or fans to games. These data were collected from various internet sources.

5.5   PRELIMINARY ANALYSIS

Before attempting to build a model to forecast the results of international soccer matches, we first present some preliminary data analysis.

5.5.1   TEAM WIN PERCENTAGE, IN AND OUT OF OWN CONFEDERATION

Each soccer team has its own win/draw/loss percentage. One would expect the best teams to win a higher percentage of their games, and the weaker teams to have a lower win percentage. Table 5.1 shows the highest 10 win percentages for our data set.

One would not be surprised to see most of these teams in the top 10. Two exceptions may be Australia and Saudi Arabia. The pre-2006 World Cup winning probability-odds[¶] for all the teams in Table 5.1, except Australia and Saudi Ara-

---

[†]www.staff.city.ac.uk/r.j.gerrard/football/aifr21_1.htm.

[‡]www.RSSSF.com.

[§]www.fifa.com.

[¶]Data used were for Ladbrokes bookmakers. Source: odds.bestbetting.com/world-cup-2006/fifa-world-cup-germany-2006/winner.

TABLE 5.1

Top 10 win percentages by team.

Position	Team	Win percentage
1	Brazil	63.25%
2	Spain	60.86%
3	France	60.78%
4	Argentina	58.38%
5	Italy	57.74%
6	Germany	57.71%
7	Czech Republic	57.04%
8	Australia	55.46%
9	Portugal	55.22%
10	Saudi Arabia	53.98%

TABLE 5.2

Top 10 win percentages in games versus own confederation.

Position	Team	Win percentage
1	Australia	90.69%
2	New Zealand	64.28%
3	Saudi Arabia	64.02%
4	Spain	61.53%
5	Japan	61.44%
6	Mexico	61.22%
7	France	61.15%
8	Korea Republic	60.00%
9	Iran	59.37%
10	Italy	59.01%

bia, were in the range 0.289 to 0.026, while for Australia and Saudi Arabia, the probability-odds were 0.008 and 0.001, respectively.

Investigating further, if we only consider matches where teams play against teams from their own confederation, the analysis reveals some interesting findings, shown in Table 5.2.

Australia wins over 90% of matches when playing against teams in its own confederation. This may explain why the country is eighth in the overall win percentage list. A similar argument can be used to explain Saudi Arabia's presence in the overall top ten. What are these tables telling us though? Are Australia and Saudi Arabia simply much stronger teams than the others in their confederation, or are these results merely representing the fact that each of their confederations are not as strong as the other confederations? Is this phenomenon dealt with appropriately in the FIFA rankings?

5.5.2 INTERNATIONAL SOCCER VERSUS DOMESTIC SOCCER

Home advantage is a well-known phenomenon in soccer (see, for example, Clarke and Norman, 1995). If one were told that team i was playing at home against team j, and no other information was known then the best forecast we could make would be to predict a home win. Home advantage is often analyzed by calculating the proportion of home wins, draws, and away wins. These statistics are often quoted when investigating the competitive balance of domestic soccer leagues (see, for example, Koning, 2000). It is also a good benchmark for forecasters when modelling match results. When discussing the performance of soccer tipsters, Forrest and Simmons (2000), comment that newspaper editors would do well to find a tipster who would outperform a simple strategy of always forecasting a home team win. The percentage of home team wins, draws, and away wins for our data set on international soccer during the period under consideration is 50.3% (3385), 24.6% (1656), and 25.1% (1689) respectively. For comparison purposes the corresponding figures for the English Premier League over the same period are 46.8% (1424), 26.1% (794), and 27.0% (822). At first glance it would seem that the home win percentage is higher for international football than for top flight domestic football in England. To check this statistically we perform two hypothesis tests for differences in the percentage of (a) home wins in domestic and international football, and (b) away wins in domestic and international football. Our null and alternative hypotheses are given by: $H_0 : p_d - p_i = 0$ versus $H_1 : p_d - p_i < 0$. The test statistic is given by

$$Z = \frac{\hat{p}_d - \hat{p}_i}{\sqrt{\dfrac{\hat{p}_d(1 - \hat{p}_d)}{n_d} + \dfrac{\hat{p}_i(1 - p_i)}{n_i}}}$$

where p_d and p_i are the proportions of home/away wins for domestic and international soccer respectively, n_d and n_i are the total number of matches played at non-neutral venues for domestic and international soccer respectively. The resultant p-values are 0.001 for the hypothesis test on home wins and 0.022 for away wins, and we reject equality in both cases. Such a result suggests that home advantage is larger for international soccer than for domestic soccer, as measured by both the number of home wins and the number of away wins. Why might this be? It may be that since teams and fans are travelling greater distances, the effect of being at home is magnified from domestic soccer. Indeed we see later that the difference in distance travelled by the two teams is a significant variable in a forecasting model for match result.

Within a match, the relationship between each team's number of goals is an interesting aspect of soccer to study. For domestic soccer various authors have investigated such issues. The results suggest that goals by the two teams display only slight positive or no correlation, see for example, Reep and Benjamin (1968). For international soccer, goals are significantly negatively correlated, see Table 5.3 (all figures are statistically significant at the 99% level).

This is a striking result. Why does such a difference in the relationship exist? We offer one possible explanation. It may be that there is a structural difference

TABLE 5.3
Correlation for goals scored by each team in a match.

Correlation	$\text{cor}(g_1, g_2)$
Pearson	−0.194
Kendall's Tau	−0.154
Spearman Rho	−0.185

TABLE 5.4
Correlation for goals scored by each team in a match.

Division	$\text{cor}(g_1, g_2)$
1-20	0.032
21-40	−0.021
41-60	0.101
61-80	−0.009
81-100	0.049

in competitive balance between the two forms of the game. For domestic soccer, the teams are separated into divisions; with the top division having the twenty (or so) top teams in the country, the second division having the next best twenty teams and so on. The vast majority of matches are between the twenty teams from each division all playing each other, and thus the matches are closely competed. Thus, the two relatively equally matched teams will tend to score similar numbers of goals, with the total number of goals scored being dependent more on the tempo and style of the match, rather than the relative quality of the teams. National team soccer on the other hand is organized quite differently. The majority of the matches are qualifying games for major tournaments, and teams are given a seeding so as to avoid two top teams meeting in the early stages of a competition. The resulting effect may be that a large proportion of the games have low competitive balance, causing a negative relationship between the two teams' scores.

A simple way to test whether the negative correlation between goals scored by each team in a match for international soccer matches is caused by low competitive balance would be to look at the correlation for international matches involving a certain division of the teams. As for domestic soccer, the divisions contain similarly able teams, with the first division having only the top 20 teams included; the next division have the next 20 ranked teams in and so on. Table 5.4 shows the Pearson correlation coefficients for goals by each team in a match played between teams in the same division.

The correlations given in Table 5.4 are not statistically significantly different from zero, suggesting that as conjectured, when the two teams are more closely matched in terms of ability, and the competitive balance is higher, then the correlation between the two teams goals scored is as for domestic soccer.

5.6   FORECASTING SOCCER MATCHES

To date, much of the literature on forecasting match outcome has been related to domestic soccer leagues. The forecasting models for such data used can loosely be allocated to two categories:

1. directly forecast the match result, i.e., the probabilities of win, loss, or a draw, using an ordinal regression model such as ordered logit or ordered probit. This is often the approach taken by economists, see, for example, Forrest, Goddard, and Simmons (2005);

2. forecast the probabilities of each possible number of goals scored by each team and then infer result probabilities. A bivariate Poisson regression model, or some variation of this model, such as a diagonally inflated bivariate Poisson model, is often employed. This is the approach often used by statisticians, see for example, Karlis and Ntzoufras (2003).

Evidence suggests very little difference in performance of each approach in modelling match outcome, see, for example, Goddard (2005). Thus, we could approach the problem using either method. However, as already discussed, one key difference between domestic soccer and national team soccer exists which renders the bivariate models used in the literature to date inapplicable in this case.

As a consequence of the bivariate Poisson model of Griffiths and Milne (1978), the two dependent variables must have non-negative correlation. This can be seen if one interprets the bivariate Poisson model as the product of three univariate Poisson probability functions. The joint probability function is given by

$$\Pr(X = x, Y = y) = \exp(-\lambda_1 - \lambda_2 - \lambda_3) \sum_{k=0}^{\min(x,y)} \frac{\lambda_1 \lambda_2 \lambda_3}{(x-k)!(y-k)!k!}.$$

It can be shown that $\mathrm{cov}(X, Y) = \lambda_3$, and must therefore be positive since λ_3 is a Poisson rate. For domestic soccer this model may be applicable, whereas for international soccer, since there is a negative relationship in the two teams' goals, the bivariate Poisson model used in previous studies cannot be employed. Thus, in order to forecast international soccer match results we employ an ordinal regression model. It should be noted that bivariate Poisson related models can be generated with the use of copulas, see, for example, McHale and Scarf (2006).

5.7   USING THE FIFA WORLD RANKINGS TO FORECAST MATCH RESULTS

In this section we present the results of using the FIFA world rankings to forecast match results with an ordered probit regression model. In addition, we use information from past results to try to improve the forecasts made and assess whether the world rankings have already taken such information into account. The variables used in this chapter are as follows:

- The dependent variable is match result, 1 for a win, 0 for a draw, and −1 for a loss.

Several covariates, and interactions between them, have been considered:

- Home/away variable, set equal to 1 if the team is at home, 0 if the game is played on neutral ground, and −1 if the team is away from home. Within a match the teams have been ordered alphabetically;

- Distance from capital city to game location;

- Difference in world ranking position between the two teams;

- Change in world rankings for each team during the previous 12 months;

- Dummy variables for the type of match, namely: major tournament (World Cup or confederation championship), minor tournament (other FIFA sanctioned tournament), qualifier or friendly;

- Dummy variables for the confederation of each team and dummy variables if the teams are from different confederations;

- Past match results. For each game we have obtained the previous twenty results for each team. In addition to the results, we have obtained the world rankings for each of the twenty opposition teams at the time of the game.

5.7.1 REACTION TO NEW INFORMATION

First, we assess how the FIFA world rankings react to recent information. Specifically we fit an ordered probit model to forecast match result and use a likelihood ratio test to test whether or not the movement in rankings of each team during the previous 12 months is a significant covariate in the model.

To assess the performance of the world rankings fairly, we have added variables that would not be expected to have been incorporated into the rankings. For example, we have included a home dummy variable, a difference in distance travelled proxy, and interaction variables between the ranking difference and the type of match: friendly, qualifier, or minor tournament, with major tournament being the reference category. The results, shown in Table 5.5, confirm that the rankings, if forecasting match results is the objective, do not react quickly enough to recent information since the changes in rankings over the past 12 months for both teams are significant variables in the regression model. A likelihood ratio test between this model and one with coefficients on the rankings change variables constrained to be zero produces a test statistic of 29.07 compared to a critical value, $\chi^2_{2,0.99}$ of 13.82. The constraints on change in ranking variables are therefore rejected, confirming that the changes in rankings variables do indeed add information to the forecasting model.

TABLE 5.5
New information regression results.

Parameter	estimate	s.e.	z
homedummy	0.33602	0.020	16.700
distdiff	−0.00003	0.000	−3.920
Δ t1rank	−0.00401	0.001	−4.240
Δ t2rank	0.00376	0.001	3.980
Rankdiff×friendly	0.00416	0.001	2.800
Rankdiff×qualifier	0.00032	0.001	0.220
Rankdiff×minor	0.00348	0.002	2.280
Rankdiff	−0.01429	0.001	−10.370
cut1	−0.425	0.016	
cut2	0.339	0.015	

An interesting result is related to the coefficient estimates for the interaction variables. Since major tournament is the reference category, it can be seen that whether or not the match is a qualifying match for a major tournament has no statistically significant effect on the probability of a higher ranked team beating a lower ranked team (rankdiff negative). However, if the match is a minor tournament or, to a greater extent a friendly match, if the rankdiff is positive (a lesser team plays a better team) then the probability of the lower ranked team winning is higher than if the match was part of a major tournament. Conversely, the better team would have a lower probability of winning. This result would seem to be evidence of an often speculated hypothesis, namely, that better teams try harder in major tournaments and fewer surprises occur, as opposed to friendly matches.

5.7.2   A FORECASTING MODEL FOR MATCH RESULT USING PAST RESULTS

Here we test whether the world rankings utilize the information on past results efficiently. In addition to the variables used in the last section, we now add past results as covariates in the regression model.

We will be using past results to forecast new results. It is clear that a 1-0 win for a team ranked 200 versus a team ranked 4 represents a better performance than if the teams were ranked 200 and 198, respectively. As a consequence, the past results must not only be weighted relative to the result alone but also be weighted relative to the opposition in that past match too. The parameterization of the model is crucial to getting a model that captures this intuitive idea, and a metric which measures this relationship is required. The past performance metric (ppm) proposed here is given by

$$
ppm(result) = \begin{cases} 1 + \dfrac{rankdiff - 204}{408} & \text{if } result = win, \\[2mm] 0.5 + \dfrac{rankdiff - 204}{408} & \text{if } result = draw, \\[2mm] 0 + \dfrac{rankdiff - 204}{408} & \text{if } result = loss. \end{cases}
$$

If one considers a scale of -1 to 1 for the *ppm*, -1 would represent the worst result possible, i.e., the top ranked team losing to the bottom ranked team, and 1 would represent the worst ranked team beating the best ranked team. This *ppm* is characterized by such an intuitively pleasing property. 204 is the maximum value of the ranking of any team during the period under consideration and by dividing the second term by 408 the measure is constrained to the range -1 to 1. Several other metrics were tried, and this one gave the best results. The past performance metric is plotted in Figure 5.1.

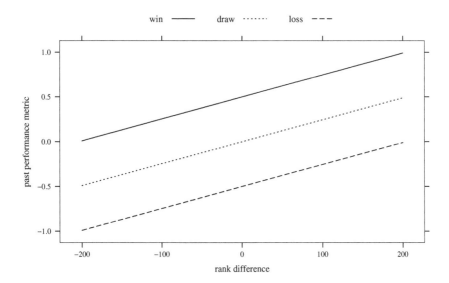

FIGURE 5.1

Past performance metric.

To test the ability of the FIFA world rankings to forecast match results we fit two models. First we fit an ordered probit model for result with the world ranking difference as the primary covariate, plus other covariates that contain information which is not meant to be reflected in the rankings, for example, a home indicator variable. The regression results are given in Table 5.6.

Next, we fit a model including the past performance metric for each of the 8 previous games. The results are shown in Table 5.7.

A likelihood ratio test between the two models gives a test statistic of 92.4, whilst the critical value at the 99.9% level, $\chi^2_{16,0.999}$, is 39.3 indicating that the past performance data add information to the forecasting model, and hence, the world rankings are not fully processing the past results of each team.

TABLE 5.6
Ordered probit estimation results.

Parameter	estimate	z
homedummy	0.310	13.50
distdiff	−0.000030	−3.86
rankdiff	−0.013	−8.45
Δt1rank	−0.003	−2.92
Δt2rank	0.003	2.93
rankdiff×friendly	0.003	1.82
rankdiff×qualifier	−0.001	−0.51
rankdiff×minor	0.004	2.34
cut1	−0.434	
cut2	0.349	

TABLE 5.7
Estimation results for full forecasting model.

Parameter	estimate	z
homedummy	0.315	13.66
distdiff	−0.000027	−3.48
rankdiff	−0.013	−8.2
Δ t1rank	−0.006	−5.19
Δ t2rank	0.005	3.94
rankdiff×friendly	0.003	1.95
rankdiff×qualifier	0.000	−0.3
rankdiff×minor	0.004	2.24
t1pp.metric1	0.108	2.81
t2pp.metric1	−0.124	−3.2
t1pp.metric2	0.097	2.5
t2pp.metric2	−0.081	−2.09
t1pp.metric3	0.065	1.66
t2pp.metric3	−0.104	−2.67
t1pp.metric4	0.165	4.22
t2pp.metric4	0.005	0.14
t1pp.metric5	0.181	4.6
t2pp.metric5	−0.024	−0.62
t1pp.metric6	0.001	0.03
t2pp.metric6	−0.031	−0.780
t1pp.metric7	0.069	1.750
t2pp.metric7	−0.029	−0.740
t1pp.metric8	0.079	2.030
t2pp.metric8	−0.066	−1.690
cut1	−0.430	
cut2	0.362	

In the model we use a lag of eight games to input past scoring records. More lags were tried, however, the forecasting performance, as measured by number of correct results inferred did not change significantly as more lags were added. In addition to this, it is interesting to note that across our data set, on average a national team plays approximately eight matches in any one calendar year. Thus using eight lags can be rationalised in this manner — eight matches corresponds to approximately one season of football.

5.8 CONCLUSION

In this chapter we have investigated whether the FIFA world rankings list, as calculated between 1998 and 2006, uses information on past results efficiently, as judged by its efficacy in a forecasting model. Using a series of ordered probit regression models, we found that the world rankings list did not use the information on past results efficiently and that FIFA may have been correct to change the way the rankings are calculated. In testing we used a past performance metric to take account of opposition quality in previous games, and it was found that this variable was statistically significant in a forecasting model for match result, which also included the world rankings difference between the two teams.

A natural way forward for this work would be to seek a rankings system that did use information on past results efficiently, and could be used to produce a rankings list free from subjectivity.

REFERENCES

Clarke, S.R. and J.M. Norman (1995). Home ground advantage of individual clubs in English soccer. *The Statistician 44*(4), 509–521.

Forrest, D., J. Goddard, and R. Simmons (2005). Odds setters as forecasters. *International Journal of Forecasting 21*, 551–564.

Forrest, D. and R. Simmons (2000). Making up the results: the work of the Football Pools Panel, 1963-1997. *The Statistican 49*(2), 253–260.

Goddard, J. (2005). Regression models for forecasting goals and match results in association football. *International Journal of Forecasting 21*, 331–340.

Griffiths, R.C. and R.K. Milne (1978). A class of bivariate Poisson processes. *Journal of Multivariate Analysis 8*, 380–395.

Karlis, D. and I. Ntzoufras (2003). Analysis of sports data by using bivariate Poisson models. *The Statistician 52*, 381–393.

Koning, R.H. (2000). Balance in competition in Dutch soccer. *The Statistician 49*(3), 419–431.

McHale, I.G. and P.A. Scarf (2006). Modelling soccer matches using bivariate discrete outcomes. Technical report, University of Salford, Salford, United Kingdom.

Reep, R. and B. Benjamin (1968). Skill and chance in Association Football. *Journal of the Royal Statistical Society A 131*, 581–585.

6

Forecasting scores and results and testing the efficiency of the fixed-odds betting market in Scottish league football

Stephen Dobson
University of Otago

John Goddard
University of Wales, Bangor

ABSTRACT

This chapter describes an analysis of the efficiency of the fixed-odds betting market for Scottish league football. Comparisons are drawn between the forecasting performance of goals-based and results-based regression models. The results-based model would have out-performed the goals-based model marginally, if it had been used to select bets during the three Scottish football seasons from 2000-1 to 2002-3. The forecasting model's selectivity capability would have eliminated most, but not all, of the effect of the bookmakers' over-round on the bettor's expected return. However, a strategy of always backing short-odds favorites would have dominated a strategy of selecting bets offering a favorable expected return according to the forecasting model. The fact that it would have been relatively straightforward to formulate a consistently profitable betting strategy in this manner suggests the standard conditions for betting market efficiency are not satisfied.

6.1 INTRODUCTION

There are two distinct strands of empirical literature on modelling the outcomes of matches in association football (soccer). The first approach involves modelling the numbers of goals scored and conceded directly. Forecasts of win-draw-lose match results can be derived indirectly, by aggregating the estimated probabilities assigned to appropriate permutations of goals scored and conceded by the two teams. The second approach involves modelling win-draw-lose results directly, using discrete choice regression models such as ordered probit or logit. Using data on English league football, Goddard (2005) has recently drawn comparisons between the forecasting performance of statistical models based on each of these two approaches. The models are estimated using the same data, and are used to produce forecasts for the same set of matches. The difference in forecasting performance between the two types of models turns out to be rather small. This finding may explain why both goals-based and results-based models have been used in the recent applied statistics and econometrics literatures, with neither of these two approaches seeming to dominate the other. In this chapter, we present a similar analysis based on data on Scottish league football.

Accurate forecasting is important primarily to bookmakers and bettors, who may have a significant financial interest in being able to assess accurately prior probabilities for football match results. Research concerning the relationship between the odds set by bookmakers for bets on the outcomes of sporting contests, and the probabilities associated with these outcomes, forms a subfield within the literature on financial market efficiency. Although much of this literature focuses on racetrack betting, the market for fixed-odds betting on match outcomes in professional football has also attracted some attention. A divergence between the bookmakers' odds and the true probabilities implies a violation of the condition for the efficiency of the fixed-odds betting market, and creates opportunities for sophisticated bettors to extrapolate from historical data on match outcomes (or other relevant information), in order to formulate a betting strategy that yields either a positive return, or (at least) a loss smaller than would be expected by an unsophisticated or indiscriminate bettor due to the loading contained in the odds that allows for the bookmaker's costs and profit. Much of the literature on the efficiency of the fixed-odds betting market focuses on betting on English football. This chapter extends the range of evidence that is available in this literature, by presenting for the first time an analysis of the efficiency of the fixed-odds betting market for Scottish league football.

This chapter is structured as follows. Section 6.2 reviews the previous academic literature on modelling goal scoring and match results in association football (soccer). Section 6.3 specifies the models that are estimated in this study. Section 6.4 describes the Scottish league football match results data set, and presents and interprets the estimation results for the goal-based and results-based models. Section 6.5 presents an analysis of the efficiency of the market for fixed-odds betting on Scottish league football. Finally, Section 6.6 summarizes and concludes.

6.2   LITERATURE REVIEW

In an early contribution to the literature on modelling goal scoring in football, Maher (1982) develops a model in which the home and away team scores follow independent Poisson distributions, with means which are the product of parameters reflecting the attacking and defensive capabilities of the two teams. The model does not predict scores or results ex ante. Although goodness-of-fit tests show that the model provides a reasonably accurate approximation to the data, separate examination of the observed and expected distributions of the difference between the scores of the two teams reveals a tendency to underestimate the proportion of drawn matches. This is attributed to interdependence between the scores of the home and away teams, and is corrected by modelling scores using a bivariate Poisson distribution.

In fixed-odds betting on Scottish (and English) football, the betting odds (the prices at which bets are accepted) are fixed by the bookmakers several days before the match. The odds are not adjusted as bets are placed, even if new information is received that is relevant to forecasting the match result. Traditionally, bets were placed over the counter in betting shops located in the high streets of most UK towns and cities. Although an increasing proportion of betting now takes place online, and new online betting exchanges compete directly with the high-street bookmakers for business, the old-fashioned high-street betting shops continue to operate.

In the first analysis of the efficiency of the fixed-odds betting market for English football, Pope and Peel (1989) estimate least squares and logit regressions of match results on bookmaker odds, and evaluate directly the mean returns on bets placed on home wins, draws, and away wins at short, medium, and long odds (i.e., bets on high probability, medium probability, and low probability outcomes, respectively), using data from the 1981-2 season. There is some evidence of divergences between the bookmakers' odds and match outcomes, but the divergences are insufficient to translate into betting strategies capable of producing a positive post-tax return. In a more recent study based on similar methodology, Cain, Law, and Peel (2000) estimate negative binomial regressions for goals scored using exact score betting odds as covariates, with data from the 1991-2 season. In a holdout sample comprising more than 800 matches from the same season, bets at short odds are found to offer a higher average return than bets at long odds. Similarly, exact score bets on low scores offer higher average returns than bets on high or unusual scores. These results are interpreted as evidence of favourite-longshot bias.

Dixon and Coles (1997) develop a forecasting model capable of generating ex ante probabilities for goals and match results. Home and away team goals follow univariate Poisson distributions, and for low-scoring matches an ad hoc adjustment to the probabilities corrects for interdependence. Using data on 6629 league and cup matches played between seasons 1992-3 to 1994-5, the estimates for each team's attacking and defensive capabilities are updated continuously. Using betting odds data from the 1995-6 season from more than one high-street bookmaker, a betting strategy that involves selecting bets for which the ratio of the model probabilities to the bookmaker probabilities exceeds a given value is found to be profitable. Dixon and Pope (2004) compare probabilistic forecasts obtained from the Dixon-Coles model with probabilities inferred from UK bookmakers' prices for fixed-odds betting. It is found that correct scores betting offers more opportunities than match results betting for detecting mispriced bets, and for the development of profitable betting strategies.

Using a forecasting approach similar to that of Dixon-Coles, Rue and Salvesen (2000) allow the attacking and defensive parameters for all teams to vary randomly over time. The estimates of these parameters are updated as new data on match outcomes are obtained. The estimation includes 1684 matches from English football's top division and 2208 matches from the second division, played during seasons 1993-4 to 1996-7 inclusive. Bets on 112 matches from the 1997-8 season, selected by comparing the model probabilities with the bookmaker's odds, are found to produce a positive return. Subsequently, Crowder, Dixon, Ledford, and Robinson (2002) have proposed a procedure for updating the team strength parameters that is computationally less demanding than the one used originally by Rue and Salvesen.

Statistical models for match results based on ordered probit or logit regression have been used by Koning (2000), Kuypers (2000), Goddard and Asimakopoulos (2004), and Forrest, Goddard, and Simmons (2005). As part of an analysis of changes in competitive balance in Dutch football, Koning (2000) advocates modelling results directly, rather than indirectly through scores, partly on grounds of simplicity: fewer parameters are required, the estimation procedures are more straightforward, and the specified ordered probit model lends itself quite easily to the inclusion of dynamics or other explanatory variables. Furthermore, the problem of interdependence between the home and away team scores is finessed using this approach. Kuypers (2000) fits an ordered probit model to results data from English 1733 league matches played during the 1993-4 season. A betting strategy based on the computation of expected returns is found to be profitable when applied to a holdout sample comprising 1649 league matches from the 1994-5 season.

Goddard and Asimakopoulos (2004) estimate an ordered probit match results forecasting model using covariates generated from past match results and other data. The estimation period comprises all league matches during a ten-season window prior to current season: accordingly, the sample size for each estimation is around 19000 matches. In a holdout sample comprising 3139 league matches from the 1998-9 and 1999-2000 seasons, bets placed on the match outcome with the highest expected return produce a positive return over the final three months of both seasons. Forrest et al. (2005) use a modified version of the Goddard and Asimakopoulos model to compare the information content and forecasting capability of bookmakers' odds and match result probabilities derived from the model. The holdout sample, comprising 9729 league matches played during seasons 1998-9 to 2002-3 (inclusive) and odds posted by five different high-street bookmakers, is considerably larger than in any previous study. It is found that the superiority of forecasting model relative to the bookmakers' odds diminished during holdout sample period. Combining information derived from the bookmakers' odds and model probabilities produces better forecasts than either information set individually.

The impact of specific factors on match outcomes has been considered in several other studies. Barnett and Hilditch (1993) investigate whether artificial playing surfaces, adopted by a few teams during the 1980s and 1990s, conferred an additional home-team advantage. Ridder, Cramer, and Hopstaken (1994) show that player dismissals have a negative effect on the match outcome for the teams concerned. Clarke and Norman (1995) quantify the effect of home advantage on match outcomes. Dixon and Robinson (1998) investigate variations in the scoring rates of the home and away teams during the course of a match. The scoring rates at any time are partly dependent on the time elapsed, and partly on which (if either) team is leading at the time. Forrest and Simmons (2000a,b) investigate the predictive quality of newspaper tipsters' results forecasts, and the performance of the pools panel in providing hypothetical results for postponed matches. Audas, Dobson, and Goddard (2002) examine whether managerial change has any short-term impact on subsequent team performance measured by match results. Dyte and Clarke (2000) examine the relationship between a set of pretournament national team rankings, and the same teams' performance in the 1998 World Cup tournament. Torgler (2004) presents a similar analysis for the 2002 World Cup, and Goddard and Thomas (2006) examine the efficiency of the fixed-odds betting market for matches played during the Euro 2004 (European Championship) tournament.

6.3 REGRESSION MODELS FOR GOAL SCORING AND MATCH RESULTS

Section 6.3 specifies the bivariate Poisson and ordered probit regression models that are used in the present study to produce forecasts of match results for Scottish league football matches. For further discussion of the model specifications, see Goddard (2005). Definitions of the covariates that are included in these models are as follows:

$F_{iys}^d = f_{iys}^d/n_{iy}$, where f_{iys}^d = home team i's total goals scored in matches played 0-12 months ($y = 0$) or 12-24 months ($y = 1$) before current match; within the current season ($s = 0$) or previous season ($s = 1$) or two seasons ago ($s = 2$); in the team's current division ($d = 0$) or one ($d = \pm 1$) or two ($d = \pm 2$) divisions above or below the current division; and $n_{iy} = i$'s total matches played 0-12 months ($y = 0$) or 12-24 months ($y = 1$) before current match.

$A_{iys}^d = a_{iys}^d/n_{iy}$, where a_{iys}^d = home team i's total goals conceded, defined for the same y, s, d as above; n_{iy} defined as above.

$P_{iys}^d = p_{iys}^d/n_{iy}$, where p_{iys}^d is i's total "points" score, on a scale of 1=win, 0.5=draw, 0=loss, defined for the same y, s, d as above; n_{iy} defined as above.

S_{im}^H = goals scored in mth most recent home match by i.

C_{im}^H = goals conceded in mth most recent home match by i.

S_{in}^A, C_{in}^A = goals scored and conceded in nth most recent away match by i.

R_{im}^H = Result (1=win, 0.5=draw, 0=loss) of i's mth most recent home match.

R_{in}^A = Result of i's nth most recent away match.

$SIGH_{ij}$ = 1 if match is important for championship, promotion, or relegation issues for i but not for away team j; 0 otherwise.

$SIGA_{ij}$ = 1 if match is important for j but not for i; 0 otherwise.

$DIST_{ij}$ = Geographical distance between the grounds of i and j.

For convenience, the following notation is used below to refer to groups of covariates:

$x_A = \{F_{iys}^d, A_{jys}^d, S_{im}^H, S_{in}^A, C_{jn}^H, C_{jm}^A\}$: home team attack, away team defense goals covariates;

$x_B = \{A_{iys}^d, F_{jys}^d, C_{im}^H, C_{in}^A, S_{jn}^H, S_{jm}^A\}$: home team defense, away team attack goals covariates;

$x_C = \{P^d_{iys}, P^d_{jys}, R^H_{im}, R^A_{in}, R^H_{jn}, R^A_{jm}\}$: home and away team results covariates;

$z = \{SIGH_{ij}, SIGA_{ij}, DIST_{ij}\}$: other covariates.

In the specification of the bivariate Poisson regression with goals scored and conceded as the dependent variables, S^H_{i0} $(=C^A_{j0})$ and C^H_{i0} $(= S^A_{j0})$ denote the goals scored and conceded by home team i in the current match, respectively. For reasons that are discussed below, in this study a copula is used to construct the bivariate Poisson distribution that is used in the goals-based model. Let $f_1(s) = \Pr(S^H_{i0} = s)$ and $f_2(c) = \Pr(C^H_{i0} = c)$ for $s, c = 0, 1, 2, \ldots$ denote the marginal probability functions for S^H_{i0} and C^H_{i0}, where $f_1(s) = \exp(-\lambda_{1ij})\lambda^s_{1ij}/s!$ and $f_2(c) = \exp(-\lambda_{2ij})\lambda^s_{2ij}/s!$. The joint probability distribution function is obtained by substituting these two univariate distribution functions into the Frank copula (Lee, 1999; Dawson, Dobson, Goddard, and Wilson, 2007). Let $F_1(s)$ and $F_2(c)$ denote the univariate distribution functions for S^H_{i0} and C^H_{i0} corresponding to $f_1(s)$ and $f_2(c)$. The bivariate joint distribution function is:

$$G(F_1(s), F_2(c)) = \frac{1}{\phi} \log \left(1 + \frac{(\exp(\phi F_1(s)) - 1)(\exp(\phi F_2(c)) - 1)}{\exp(\phi) - 1} \right). \quad (6.1)$$

The ancillary parameter ϕ determines the nature of any correlation between S^H_{i0} and C^H_{i0}. For $\phi < 0$ the correlation between S^H_{i0} and C^H_{i0} is positive, and for $\phi > 0$ this correlation is negative. $G(F_1(s), F_2(c))$ is undefined for $\phi = 0$, but it is conventional to write $G(F_1(s), F_2(c)) = F_1(s)F_2(c)$ in this case. The bivariate joint probability function corresponding to $G(F_1(s), F_2(c))$ is obtained iteratively, as follows:

$$\Pr(S^H_{i0} = 0, C^H_{i0} = 0) = G(F_1(0), F_2(0))$$
$$\Pr(S^H_{i0} = s, C^H_{i0} = 0) = G(F_1(s), F_2(0)) - G(F_1(s - 1), F_2(0)),$$
$$s = 1, 2, \ldots$$
$$\Pr(S^H_{i0} = 0, C^H_{i0} = c) = G(F_1(0), F_2(c)) - G(F_1(0), F_2(c - 1)),$$
$$c = 1, 2, \ldots$$
$$\Pr(S^H_{i0} = s, C^H_{i0} = c) = G(F_1(s), F_2(c)) - G(F_1(s - 1), F_2(c))$$
$$-G(F_1(s), F_2(c - 1)) + G(F_1(s - 1), F_2(c - 1))$$
$$s, c = 1, 2, \ldots$$

The construction of the bivariate Poisson distribution via the Frank copula requires some further comment. As noted above, this method allows for unrestricted (positive or negative) correlation between S^H_{i0} and C^H_{i0}, depending on the sign of the ancillary parameter ϕ. In contrast, the standard bivariate Poisson distribution, constructed by combining three random variables with univariate Poisson distributions, is capable of accommodating positive correlation only. As is reported in Section 6.4, in the present case the sample correlation between S^H_{i0} and C^H_{i0} turns out to be negative, justifying the use of the more flexible specification that is described above.

Using Italian Serie A match results data, Karlis and Ntzoufras (2003) identify a tendency for bivariate Poisson regressions to underestimate slightly the probabilities for low-scoring draws. In the present case, likelihood-ratio tests indicate that additional parameters inflating the probabilities of 0-0 and 1-1 draws are significant. Accordingly, the adjusted bivariate probability function contains two additional parameters π and θ, as follows:

$$\widetilde{\Pr}(S_{i0}^H = s, C_{i0}^H = c) = (1 - \pi)\Pr(S_{i0}^H = s, C_{i0}^H = c) + \pi\theta \qquad \{s, c\} = \{0, 0\},$$

$$\widetilde{\Pr}(S_{i0}^H = s, C_{i0}^H = c) = (1 - \pi)\Pr(S_{i0}^H = s, C_{i0}^H = c) + \pi(1 - \theta)$$

$$\{s, c\} = \{1, 1\},$$

$$\widetilde{\Pr}(S_{i0}^H = s, C_{i0}^H = c) = (1 - \pi)\Pr(S_{i0}^H = s, C_{i0}^H = c) \qquad \{s, c\} \neq \{0, 0\}, \{1, 1\}.$$

For an ordered probit regression model with win-draw-lose match results as outcomes, the result of the match between teams i and j, denoted $R_{i0}^H(= 1 - R_{j0}^A)$, depends on the unobserved or latent variable and a Normal independent and identically distributed disturbance term, ϵ_{ij}, as follows:

Home win	$R_{i0}^H = 1$	if $\mu_2 < y_{ij}^* + \epsilon_{ij}$,
Draw	$R_{i0}^H = 0.5$	if $\mu_1 < y_{ij}^* + \epsilon_{ij} < \mu_2$,
Away win	$R_{i0}^H = 0$	if $y_{ij}^* + \epsilon_{ij} < \mu_1$.

In these expressions, μ_1 and μ_2 are parameters which determine the overall proportions of home wins, draws, and away wins. As before, we assume $y_{ij}^* = g(x_C, z)$, where g is a linear function. Maximum likelihood estimation can be used to determine the coefficients on the covariates x_C and z in the function $g(\cdot)$, together with the values of μ_1 and μ_2, that most accurately describe any particular match results data set.

6.4   DATA AND ESTIMATION RESULTS

The data set for the present study comprises 11205 matches played in the Scottish league football between seasons 1988-9 and 2002-3 (inclusive). The top division of the Scottish league is known as the Scottish Premier League (SPL), and the lower divisions operate under the umbrella of the Scottish Football League (SFL). During our sample period, there were several adjustments to the competitive structure of Scottish league football. Between 1988-9 and 1990-1 there were 38 member teams in total: 10 in the SPL, and 14 in each of the First and Second Divisions of the SFL. Between 1991-2 and 1993-4, the numbers of teams in these three divisions were adjusted to 12, 12 and 14, respectively. In 1994-5 the total membership was increased to 40 teams, and the number of divisions was increased to four, with 10 teams in each division. In 2000-1 the membership of the SPL was increased from 10 to 12, and two new teams were admitted to the SFL, bringing the combined membership to its present complement of 42 teams. The three SFL divisions have continued to operate with a membership of 10 teams per division. In all divisions, the teams play one another either three times or four times per season.

Scottish league football is notorious for the asymmetry that exists between the popularity, wealth, and success of the two leading teams, Celtic and Rangers, and the

TABLE 6.1

Percentages of home wins, draws, and away wins, average goals scored by
home and away teams, SPL and SFL, 1988-9 to 2002-3.

Season	Matches	Home (%)	Draw (%)	Away (%)	Home goals per match	Away goals per match
1988-9	726	43.9	28.8	27.3	1.46	1.22
1989-90	726	44.1	29.2	26.7	1.53	1.19
1990-1	726	42.4	26.6	31.0	1.47	1.20
1991-2	801	40.0	26.3	33.7	1.45	1.26
1992-3	801	41.4	26.0	32.6	1.51	1.29
1993-4	801	38.6	31.2	30.2	1.37	1.14
1994-5	720	42.8	26.1	31.1	1.47	1.18
1995-6	720	40.4	25.6	34.0	1.45	1.25
1996-7	720	42.5	25.6	31.9	1.45	1.26
1997-8	720	40.7	26.3	33.1	1.50	1.30
1998-9	720	41.1	23.9	35.0	1.44	1.30
1999-2000	720	41.3	24.9	33.9	1.52	1.29
2000-1	768	40.2	25.3	34.5	1.47	1.24
2001-2	768	44.4	24.9	30.7	1.52	1.25
2002-3	768	42.1	25.0	32.9	1.55	1.30

rest. Not since 1984-5 has any team other than Celtic or Rangers won the SPL. At the
opposite end of the scale, most teams in the lower divisions of the SFL operate on a
very small scale indeed, with average match attendances in the low hundreds in most
cases. Overall there is greater competitive imbalance among the 42 member teams of
the SPL and SFL than there is among the 92 member teams of the English Premier
League and Football League. Even within the divisions of the SPL and SFL, the
level of competitive imbalance is reflected in a variance of match result probabilities
or bookmaker odds for individual matches that tends to be larger than the variance
for English league matches.

Table 6.1 and Table 6.2 present summary data for the percentages of matches that
finished as home wins, draws, and away wins by season; the average numbers of
goals scored by the home and away teams per match by season; and the bivariate
percentage distribution for the numbers of goals scored by the home and away teams
per match over the entire sample period. In common with the data for several other
countries, it appears there was some reduction in the importance of home advantage
over this period, although the trend is not particularly strong. The introduction of
three league points for a win (rather than two), which took place in Scotland from
the start of the 1995-6 season, appears to have produced a small increase in the
average rate of goal scoring by the away team, and a reduction in the proportion of
matches finishing in draws. In comparison with similar tables compiled by Dobson
and Goddard (2001) for England, the proportions of matches finishing as home wins
and draws are lower in Scotland, and the proportion of matches finishing as away
wins is higher. Average goal scoring rates by Scottish home teams are similar to those

TABLE 6.2

Joint and marginal distributions of goals scored by home and away teams, SPL and SFL, 1988-9 to 2002-3.

Home goals	Away goals							Total
	0	1	2	3	4	5	6+	
0	7.38	7.90	5.02	2.71	0.85	0.19	0.19	24.24
1	9.30	12.54	7.08	3.07	1.20	0.44	0.12	33.74
2	7.19	8.02	5.45	1.76	0.60	0.19	0.04	23.25
3	3.55	3.97	2.38	0.87	0.46	0.04	0.04	11.32
4	1.62	1.87	0.90	0.28	0.12	0.03	0.00	4.82
5	0.65	0.71	0.26	0.12	0.04	0.01	0.02	1.79
6+	0.39	0.29	0.12	0.03	0.00	0.01	0.00	0.84
Total	30.09	35.29	21.22	8.84	3.28	0.89	0.39	100.00

in England, but average goal scoring rates by Scottish away teams are significantly higher. It is interesting to note that in our Scottish sample, the correlation between the goals scored in each match by the home team and the away team is negative, with a value of -0.0677. This is in contrast to the positive correlation that is usually obtained from English league match results data.

The bivariate Poisson goals-based model and the ordered probit regression results-based model specified in Section 6.3 are used to generate two sets of win-draw-lose match results forecasts for matches played in the Scottish football league for each of the three seasons, 2000-1, 2001-2, and 2002-3. The forecasts take the form of ex ante home win, draw, and away win probabilities. The two sets of forecasts for each season are obtained from versions of the models estimated using data for the preceding 10 seasons. Accordingly, the estimation period for the models used to obtain the forecasts for 2000-1 is 1990-1 to 1999-2000; the estimation period for the 2001-2 forecasts is 1991-2 to 2000-1; and the estimation period for the 2002-3 forecasts is 1992-3 to 2001-2. Tables 6.3 and 6.4 report the estimated versions of the two models, based on data for seasons 1992-3 to 2001-2 (inclusive). The contribution of each set of covariates is now considered.

In the bivariate Poisson model, the average goals scored and goals conceded variables F_{iys}^d and A_{iys}^d (for team i and their counterparts for team j), calculated over the 24 months prior to the current match, are the main indicators of attacking and defensive capability, or team quality. The average "points per match" (or win ratio) covariates P_{iys}^d and P_{jys}^d play an equivalent role in the ordered probit regression model. The indexing of these variables allows for separate contributions to the team quality measures from goals scored and conceded in matches played: 0-12 months ($y = 0$) or 12-24 months ($y = 1$) before the current match; within the current season ($s = 0$) or previous season ($s = 1$) or two seasons ago ($s = 2$); and in the team's current division ($d = 0$) or one ($d = \pm1$) or two ($d = \pm2$) divisions above or below the current division. The estimated coefficients on all of these variables are predominantly correctly signed and well defined. Preliminary experimentation indicated that

TABLE 6.3
Estimation results: bivariate Poisson regression for goals 1992-3 to 2001-2 seasons.

	Mean home goals(λ_{1ij})				Mean away goals(λ_{2ij})		
Cov.	Coeff.	Cov.	Coeff.	Cov.	Coeff.	Cov.	Coeff.
F_{i00}^0	0.359***	A_{j00}^0	0.477***	F_{j00}^0	0.202***	A_{i00}^0	0.265***
F_{i01}^{+1}	0.407***	A_{j01}^{+1}	0.167***	F_{j01}^{+1}	0.369***	A_{i01}^{+1}	0.199***
F_{i01}^0	0.317***	A_{j01}^0	0.240***	F_{j01}^0	0.307***	A_{i01}^0	0.304***
F_{i01}^{-1}	0.143**	A_{j01}^{-1}	0.358***	F_{j01}^{-1}	0.191***	A_{i01}^{-1}	0.457***
F_{i11}^{+1}	0.175**	A_{j11}^{+1}	-0.066	F_{j11}^{+1}	0.132	A_{i11}^{+1}	0.169***
F_{i11}^0	0.086*	A_{j11}^0	-0.061	F_{j11}^0	0.076	A_{j11}^0	0.191***
F_{i11}^{-1}	0.067	A_{j11}^{-1}	-0.017	F_{j11}^{-1}	0.045	A_{i11}^{-1}	0.342***
F_{i12}^{+2}	0.638**	A_{j12}^{+2}	0.030	F_{j12}^{+2}	0.608*	A_{i12}^{+2}	0.002
F_{i12}^{+1}	0.258***	A_{j12}^{+1}	0.109**	F_{j12}^{+1}	0.287***	A_{i12}^{+1}	0.041
F_{i12}^0	0.168***	A_{j12}^0	0.180***	F_{j12}^0	0.147***	A_{i12}^0	0.081
F_{i12}^{-1}	0.133***	A_{j12}^{-1}	0.316***	F_{j12}^{-1}	0.046	A_{i12}^{-1}	0.219**
F_{i12}^{-2}	-0.270	A_{j12}^{-2}	0.075	F_{j12}^{-2}	0.124	A_{i12}^{-2}	-0.038
S_{i1}^H	0.016**	C_{j1}^H	0.008	S_{j1}^A	0.006	C_{i1}^A	0.007
S_{i2}^H	0.006	C_{j2}^H	0.028***	S_{j2}^A	0.021**	C_{i2}^A	-0.003
S_{i3}^H	0.004	C_{j3}^H	0.007	S_{j3}^A	0.024**	C_{i3}^A	0.010
S_{i4}^H	0.011	C_{j4}^H	0.003	S_{j4}^A	0.023**	C_{i4}^A	0.002
S_{i5}^H	0.009	C_{j1}^A	0.008	S_{j5}^A	0.007	C_{i1}^H	0.010
S_{i6}^H	0.000	C_{j2}^A	-0.005	S_{j6}^A	0.022**	C_{i2}^H	0.019**
S_{i7}^H	0.017**	C_{j3}^A	0.000	S_{j7}^A	0.007	C_{i3}^H	-0.008
S_{i8}^H	-0.005	C_{j4}^A	0.000	S_{j8}^A	-0.008	C_{i4}^H	0.014
S_{i9}^H	0.009	C_{j5}^A	0.000	S_{j9}^A	0.000	C_{i5}^H	0.023**
S_{i1}^A	-0.002	C_{j6}^A	0.004	S_{j1}^H	0.015*	C_{i6}^H	0.000
S_{i2}^A	0.008	C_{j7}^A	-0.002	S_{j2}^H	0.000	C_{i7}^H	0.011
S_{i3}^A	-0.002	C_{j8}^A	-0.001	S_{j3}^H	0.015*	C_{i8}^H	-0.006
S_{i4}^A	-0.003	C_{j9}^A	-0.003	S_{j4}^H	0.006	C_{i9}^H	0.000
$SIGH_{ij}$	-0.018	$DIST_{ij}$	0.000	$SIGH_{ij}$	-0.028	$DIST_{ij}$	0.000
$SIGA_{ij}$	-0.153***	const.	-0.988***	$SIGA_{ij}$	0.061	const.	-1.16***

Estimations are over 7122 match observations. *** denotes significantly different from zero, 1% level, two-tail test; ** denotes 5% level; * denotes 10% level. $\hat{\phi} = 0.154^*$, $\hat{\pi} = 0.028^{***}$, $\hat{\theta} = 0.511^{***}$.

TABLE 6.4

Estimation results: 1992-3 to 2001-2 seasons: Ordered probit regression.

Cov.	Coeff.	Cov.	Coeff.	Cov.	Coeff.	Cov.	Coeff.
P^0_{i00}	2.283***	P^0_{j00}	−0.881***	R^H_{i1}	0.011	R^H_{j1}	−0.030*
P^{+1}_{i01}	1.708***	P^{+1}_{j01}	−0.642***	R^H_{i2}	0.039**	R^H_{j2}	−0.010
P^0_{i01}	1.394***	P^0_{j01}	−0.466***	R^H_{i3}	−0.032*	R^H_{j3}	−0.025
P^{-1}_{i01}	0.843***	P^{-1}_{j01}	−0.203*	R^H_{i4}	−0.007	R^H_{j4}	0.010
P^{+1}_{i11}	0.373*	P^{+1}_{j11}	−0.349*	R^H_{i5}	0.021	R^A_{j1}	−0.030*
P^0_{i11}	0.219**	P^0_{j11}	−0.177*	R^H_{i6}	−0.024	R^A_{j2}	0.001
P^{-1}_{i11}	0.132	P^{-1}_{j11}	−0.155*	R^H_{i7}	0.015	R^A_{j3}	−0.032*
P^{+2}_{i12}	1.780***	P^{+2}_{j12}	−1.519*	R^H_{i8}	0.003	R^A_{j4}	−0.050***
P^{+1}_{i12}	0.541***	P^{+1}_{j12}	−0.653**	R^H_{i9}	0.000	R^A_{j5}	−0.010
P^0_{i12}	0.281***	P^0_{j12}	−0.278***	R^A_{i1}	0.008	R^A_{j6}	−0.004
P^{-1}_{i12}	0.153*	P^{-1}_{j12}	−0.043	R^A_{i2}	−0.003	R^A_{j7}	0.006
P^{-2}_{i12}	−0.141	P^{-2}_{j12}	−0.422	R^A_{i3}	0.023	R^A_{j8}	−0.001
$SIGH_{ij}$	−0.029	$DIST_{ij}$	0.001	R^A_{i4}	−0.011	R^A_{j9}	−0.007
$SIGA_{ij}$	−0.107*						

Estimations are over 7122 match observations. *** denotes significantly different from zero, 1% level, two-tail test; ** denotes 5% level; * denotes 10% level. $\hat{\mu}_1 = -0.367^{***}$, $\hat{\mu}_2 = 0.330^{***}$.

the coefficients on F^d_{iys}, A^d_{iys}, F^d_{jys}, A^d_{jys}, and P^d_{iys}, P^d_{jys} were highly significant for $y = 0, 1$ (data from matches played 0-12 months and 12-24 months before the current match); but not for $y = 2$ (24-36 months before the current match).

The recent goals scored and conceded variables S^H_{im}, S^A_{in}, C^H_{in}, and C^A_{im} (and their counterparts for team j) allow for the inclusion in the bivariate Poisson regression model of goals data from each team's few most recent matches. S^H_{im} is the number of goals scored by team i in its mth most recent home match; S^A_{in} is the number of goals scored by team i in its nth most recent away match; C^H_{im} and C^A_{in} are similarly defined for goals conceded by team i. The recent win-draw-lose match results variables R^H_{im} and R^A_{in} (and their counterparts for team j) play an equivalent role in the ordered probit regression model. In general, the estimated coefficients on the recent goals and results covariates tend to be rather erratic. For consistency with the English league football forecasting models developed by Goddard and Asimakopoulos (2004) and Goddard (2005), data are included from the home team's last 9 home matches and last 4 away matches, and from the away team's last 4 home matches and last 9 away matches. However, fewer of the estimated coefficients are significant in the present case than in the case of the English models reported in the earlier papers. In part, this is because the models reported here are estimated over smaller sample sizes.

The identification of matches with importance for end-of-season championship, promotion and relegation issues is relevant if match outcomes are affected by incentives: if a match is important for one team and unimportant for the other, the teams may contribute different levels of effort. A match is deemed to be important if it is possible (before the match is played) for the team in question to win the championship or be promoted or relegated, assuming all other teams currently in contention for the same end-of-season outcome take one point on average from each of their remaining fixtures. The coefficients on $SIGH_{ij}$ and $SIGA_{ij}$ should be positive and negative respectively in the equations for mean home goals (λ_{1ij}) and match results (y_{ij}^*), and negative and positive respectively in the equation for mean away goals (λ_{2ij}). The coefficients on $SIGA_{ij}$ appear to be better defined than those on $SIGH_{ij}$, and although the effects are not particularly strong, the general pattern is consistent with prior expectations.

In the case of English league football, Clarke and Norman (1995), Goddard and Asimakopoulos (2004), and Goddard (2005) have shown that geographical distance is a significant influence on match results. It has been suggested that the greater intensity of competition in local derbies may have some effect in offsetting home advantage, while the psychological or practical difficulties of long-distance travel for teams and supporters may increase home advantage in matches between teams from distant cities. In the present case, however, the estimated coefficients on $DIST_{ij}$ reported in Table 6.3 and Table 6.4 are all very small, and none is statistically significant. Perhaps this reflects the fact that the geographical configuration of football teams in Scotland is rather different from that in England. All but nine of the present complement of 42 SPL and SFL member teams are located within at most 30 miles of a straight line, running from Montrose in the northeast through the Scottish central belt to Ayr in the southwest. Such a line is only around 125 miles in length. In other words, for most Scottish fixtures travelling distances are not large, for either the teams or the spectators. Accordingly, geographical distance seems to exert less influence on Scottish match results than is the case in England.

6.5   THE EFFICIENCY OF THE MARKET FOR FIXED-ODDS BETTING ON SCOTTISH LEAGUE FOOTBALL

Section 6.5 presents comparisons between the probabilistic match result forecasts generated by the regression models described in Section 6.3 and Section 6.4, and five sets of odds posted by UK high-street bookmakers for fixed-odds betting on half-time/full-time outcomes. The odds are those compiled by the four major high-street bookmakers Coral, William Hill, Ladbrokes, and StanleyBet, and those of Super-Soccer, a specialist agency that supplies odds to small independent bookmakers. The fixed-odds betting data covers the three seasons from 2000-1 to 2002-3 inclusive. The data source is the electronic betting odds archive www.mabels-tables.com.

The bookmakers' odds are quoted in either fractional form or decimal form. With fractional odds, a quoted price of a-to-b on a home win means that if b is staked on this result, the net payoffs to the bettor are $+a$ (the bookmaker pays the winnings and returns the stake) if the bet wins; and $-b$ (the bookmaker keeps the stake) if the bet loses. With decimal odds, the quoted price for the same bet would be $(a + b)/b$. If the bet were "fair," in the sense of producing an expected return of zero to both the bookmaker and the bettor, the implicit probability would be $\theta_{ij}^H = b/(a + b)$.

A necessary condition for all three possible bets on a single match to be "fair" is $\sum_{m\in\Theta}\theta_{ij}^m = 1$ (where Θ is the set of three possible outcomes). However, in practice $\sum_{m\in\Theta}\theta_{ij}^m > 1$, because the bookmakers' odds contain margins for costs and profit. The bookmaker's over-round is $\lambda_{ij} = \sum_{m\in\Theta}\theta_{ij}^m - 1$. An implicit probability for a home win (for example) can be obtained by rescaling θ_{ij}^H as follows: $\phi_{ij}^H = \theta_{ij}^H / \sum_{m\in\Theta}\theta_{ij}^m$. By construction $\sum_{m\in\Theta}\phi_{ij}^m = 1$. ϕ_{ij}^H can be interpreted as probabilities on the assumption that the bookmakers set their odds so as to achieve the same expected return from every bet.

Different UK high-street bookmakers usually quote similar but not always identical odds for the same bet. Therefore there are arbitrage opportunities for a bettor who is willing to shop around for the bookmaker offering the most favorable (longest) odds for each bet. In this study, the efficiency tests are based on two sets of odds that are constructed from the five sets of bookmaker odds in the original data set. The first constructed set, "best odds," contains the longest odds available from any of the five sets of bookmaker odds for each possible bet. The second constructed set, "median odds," contains the third-longest (or median) odds from the five sets of bookmaker odds. "Best odds" are the odds actually available to the bettor who shops around; and "median odds" are representative of the odds available to the bettor who does not do so.

From 2304 Scottish league matches played in total during the three seasons from 2000-1 to 2002-3, the model is capable of generating predictions for 2096. The model does not generate predictions for matches involving teams that entered the league up to two calendar years before the match is played, because it requires a full set of previous match results data for both teams over a two-year period prior to the match in question. There were 22 matches for which the bookmakers' odds data were unavailable, and these matches were discarded from the sample. Therefore the final holdout sample, for which both betting odds and probabilistic forecasts are available, comprises 2074 matches. Using the scores-based bivariate Poisson model, the estimated home win, draw, and away win probabilities (i.e., win-draw-lose probabilities for the home team) are $p_{ij}^H = \sum_{s>c}\sum_c \widetilde{\Pr}(S_{i0}^H = s, C_{i0}^H = c)$, $p_{ij}^D = \sum_{s=c}\widetilde{\Pr}(S_{i0}^H = s, C_{i0}^H = c)$, and $p_{ij}^A = \sum_{s<c}\sum_{c\geq1}\widetilde{\Pr}(S_{i0}^H = s, C_{i0}^H = c)$ respectively. Using the results-based ordered probit regression model, the equivalent probabilities are $p_{ij}^H = 1 - \Phi(\hat{\mu}_2 - \hat{y}_{ij}^*)$, $p_{ij}^D = \Phi(\hat{\mu}_2 - \hat{y}_{ij}^*) - \Phi(\hat{\mu}_1 - \hat{y}_{ij}^*)$, and $p_{ij}^A = \Phi(\hat{\mu}_1 - \hat{y}_{ij}^*)$.

The pseudo-likelihood statistic, equivalent to the geometric mean of the fitted probabilities for the actual results of all matches played during the forecast period, provides a convenient summary measure of forecasting performance. The sample values of this statistic for the goals-based and results based forecasting models are very similar: $PsL = 0.3625$ for the former, and $PsL = 0.3621$ for the latter. However, there is some random, stochastic variation in the individual match result probabilities obtained from the two models. Using the methodology for investigating the efficiency of the fixed-odds betting market that is described below, the results-based model out-performs the goals-based model, despite the fact that it has a marginally lower pseudo-likelihood statistic. Accordingly, the forecasting model probabilities that are used in the analysis which follows are taken from the results-based model. A simple empirical test for the efficiency of the fixed-odds betting market takes the form of an investigation of the profitability of a simple betting strategy that selects the bet offering the highest expected return according to the probabilistic forecasts generated from the model, from the choice of three bets that are available for each match. First, the mechanics of the betting strategy are described, using as an example the six SPL fixtures that were played on the weekend of 24-25 May 2003, the final round of matches in the 2002-03 Scottish season. Then the results of applying the "highest expected return" strategy across the entire holdout sample are presented.

The application of the "highest expected return" strategy on 24-25 May 2003 is illustrated in Table 6.5. Panel 1 shows the forecasting model probabilities, p_{ij}^m, for the three possible half-time/full-time outcomes ($m \in \Theta = \{H, D, A\}$) for each of the six matches. p_{ij}^m are calculated from the ordered probit model estimated with data from seasons 1992-3 to 2001-2; $\hat{\mu}_1$ and $\hat{\mu}_2$ are the estimated values for the 2001-2 season from the same model. Panel 2 shows the "best odds" available for each of the three outcomes. Panel 3 shows the forecasting model's expected return for each bet placed at "best odds." For example, a £1 bet on a home win for Dundee United versus Aberdeen at "best odds" of 2.6 would provide a net return of +1.6 if the bet wins, and -1 if the bet loses. The expected return is $+1.6 \times 0.255 + -1 \times (1 - 0.255) = -0.337$. Panel 4 shows the selected bet for each fixture, chosen on the basis of the highest of the three expected returns, together with the match result. Finally, Panel 5 shows the returns from each of the six bets. The bettor achieves positive net returns of +1.75 and +4.5 on two of the six bets. The other four bets are unsuccessful. Across all six bets, the bettor's overall net return is +2.25.

Summary results for the application of the "highest expected return" strategy across the entire holdout sample are as follows. If one bet is placed on each of the 2074 matches at "best odds," on the outcome with the highest expected return according to the model, there are 769 winning bets in total. If £1 is staked in each bet, the overall net return is −£80.17 on a total stake of £2074. Therefore the average rate of return is $-80.17/2074 = -0.0387$, or -3.87%. For comparative purposes, it is informative to calculate the average rates of return under two alternative scenarios. In the first alternative, the same betting strategy is applied using "median odds" instead of "best odds." This produces an average return of -8.11%. In the second alternative, three bets are placed on each of the 2074 matches indiscriminately: one bet on each of the three possible outcomes, using "median odds." This produces an average return of -13.39% (slightly worse than the average of the over-rounds implicit in the five sets of bookmaker odds). Overall, a bettor who adopts the "highest expected return" strategy using "best odds" achieves a rate of return 9.52% higher than a bettor who places bets indiscriminately. Of this gain, 5.28% can be attributed to the use of the forecasting model to select bets with favorable expected returns, and 4.24% can be attributed to arbitrage between the five sets of bookmaker odds, using the "best odds" rather than the "median odds" for each bet. However, even after taking advantage of the benefits of arbitrage between the five sets of bookmaker odds, the forecasting model's selectivity capability is insufficient to fully overcome the bookmakers' over-round, or to offer the informed bettor a positive expected return.

TABLE 6.5
Outcome of highest expected return betting strategy, weekend of 24-25 May 2003.

	Dundee Utd v Aberdeen	Hibernian v Partick	Motherwell v Livingston	Hearts v Dundee	Kilmarnock v Celtic	Rangers v Dunfermline
1. Ordered probit model probabilities						
Home	0.255	0.571	0.265	0.550	0.174	0.842
Draw	0.260	0.239	0.263	0.245	0.231	0.113
Away	0.485	0.191	0.472	0.206	0.595	0.045
2. Best available bookmaker odds						
Home	2.6	1.615	2.6	1.667	13	1.1
Draw	3.25	3.6	3.4	3.75	6.5	7
Away	2.75	5.5	2.625	6.5	1.167	21
3. Expected returns						
Home	−0.337	−0.078	−0.310	−0.084	1.264	−0.074
Draw	−0.154	−0.141	−0.107	−0.082	0.500	−0.207
Away	0.333	0.048	0.239	0.336	−0.306	−0.067
4. Selected bet and match result						
	A	A	A	A	H	A
	0-2	2-3	6-2	1-0	0-4	6-1
5. Return on bet of +1						
	+1.75	+4.5	−1	−1	−1	−1

TABLE 6.6

Outcome of highest expected return and "mechanical" betting strategies, 2000-1 to 2002-3 seasons.

Decile	Bets sorted by forecasting model's expected return		Bets sorted by decimal odds		
	Mean decimal odds	Mean return	Decimal odds (range)	Number of bets	Mean return
Top	4.55	−0.165	1.1 to 1.364	156	0.072
2nd	3.11	−0.012	1.4 to 1.667	315	0.009
3rd	3.11	−0.075	1.727 to 2.2	877	−0.055
4th	3.22	0.084	2.25 to 2.5	810	−0.149
5th	3.30	−0.015	2.6 to 3.0	761	−0.058
6th	3.18	−0.145	3.1 to 3.4	533	−0.090
7th	3.22	−0.042	3.5	726	−0.050
8th	2.97	−0.152	3.6	528	−0.086
9th	3.14	−0.090	3.75 to 4.333	766	−0.029
Bottom	3.69	−0.140	4.5 to 21.0	750	−0.278

Panel 1 of Table 6.6 shows the mean returns earned by ranking all 6222 available bets (= three possible bets × 2074 matches) in the "best odds" set in descending order of their expected return evaluated using the forecasting model probabilities. For convenience, the 6222 bets are placed into ten equal-sized bands, each containing 622 or 623 bets. Panel 1 of Table 6.6 suggests that a betting rule involving the selection of bets for which the model's expected return exceeds a specific threshold would produce a disappointing performance: the bets in the highest decile by expected return turn out to produce the lowest actual return of any of the ten deciles! A possible explanation becomes apparent from inspection of the averages of the decimal odds in each of the ten deciles: the bets in the highest decile have the longest odds on average for any of the ten deciles, and therefore represent poor value for money, due to the phenomenon of favorite-longshot bias. Leaving aside the top decile, the model's selectivity capability is more clearly reflected in a generally positive relationship between expected and actual returns over the other nine deciles. However, sorting the bets by the model's expected return does not appear to provide a promising route for the formulation of a profitable betting strategy.

The results in Panel 1 of Table 6.6 suggest that a simple or "mechanical" betting strategy of always backing favorites and never backing longshots might outperform a strategy based on inspection of the forecasting model's expected returns. In Panel 2 of Table 6.6, the 6222 available bets are placed into bands defined by the length of the decimal odds. The evidence of favorite-longshot bias is very clear in Panel 2: the average returns are positive across all bets in the two shortest odds categories (1.1 to 1.364 and 1.4 to 1.667), while the longest odds band (4.5 to 21.0) produces an average return substantially lower than any of the other nine bands shown in Panel 2.

Therefore the main conclusions that emerge from Section 6.5 concerning fixed-odds betting on Scottish league football are: first, if the forecasting model is used in conjunction with a policy of arbitrage, so that the bettor always selects the best available odds, the forecasting model's selectivity capability is capable of eliminating most, but not all, of the effect of the bookmakers' over-round on the bettor's expected return; and second, a "mechanical" strategy of always backing favorites at very short odds appears to dominate a strategy of selecting bets offering a favorable expected return according to the forecasting model.

6.6 CONCLUSION

Much of the previous academic literature on the efficiency of the fixed-odds betting market focuses on betting on English football. This chapter extends the range of evidence that is available in this literature, by presenting for the first time an analysis of the efficiency of the fixed-odds betting market for Scottish league football. Scottish league football offers some interesting contrasts with its English counterpart, including fewer teams, smaller divisions with repeat fixtures, greater competitive imbalance within the league as a whole and within divisions, and a more compact geographical configuration of member teams. In the empirical analysis, comparisons have been drawn between the forecasting performance of goals-based and results-based regression models for Scottish match results. The comparisons are facilitated by estimating a set of models over data sets that are identical in all respects apart from their emphasis on goals or results, respectively.

Although the overall forecasting performance of the two types of models turns out to be very similar, it is found that the results-based model would have out-performed the goals-based model if it had been used to select bets during the three seasons from 2000-1 to 2002-3. If the results-based forecasting model had been used in conjunction with a policy of arbitrage, so that the bettor always selected the best available odds, the forecasting model's selectivity capability would have eliminated most, but not quite all, of the effect of the bookmakers' over-round on the bettor's expected return. A "mechanical" strategy of always backing favorites at very short odds would have dominated a strategy of selecting bets offering a favorable expected return according to the forecasting model. The fact that it would have been relatively straightforward to formulate a consistently profitable betting strategy in this way suggests the standard conditions for betting market efficiency are not satisfied in the case of fixed-odds betting on Scottish league football results.

REFERENCES

Audas, R., S. Dobson, and J. Goddard (2002). The impact of managerial change on team performance in professional sports. *Journal of Economics and Business 54*, 633–650.

Barnett, V. and S. Hilditch (1993). The effect of an articifial pitch surface on home team performance in football (soccer). *Journal of the Royal Statistical Society, A 156*, 39–50.

Cain, M., D. Law, and D. Peel (2000). The favourite-longshot bias and market efficiency in UK football betting. *Scottish Journal of Political Economy 47*, 25–36.

Clarke, S.R. and J.M. Norman (1995). Home ground advantage of individual clubs in English soccer. *The Statistician 44*(4), 509–521.

Crowder, M., M. Dixon, A. Ledford, and M. Robinson (2002). Dynamic modelling and prediction of English Football League matches for betting. *The Statistician 51*, 157–168.

Dawson, P., S. Dobson, J. Goddard, and J. Wilson (2007). Are football referees really biased and inconsistent? Evidence on the incidence of disciplinary sanctions in the English Premier League. *Journal of the Royal Statistical Society Series A 170*, 231–250.

Dixon, M.J. and S.C. Coles (1997). Modelling association football scores and inefficiencies in the football betting market. *Applied Statistics 46*, 265–280.

Dixon, M.J. and P.F. Pope (2004). The value of statistical forecasts in the UK association football betting market. *International Journal of Forecasting 20*, 697–711.

Dixon, M.J. and M.E. Robinson (1998). A birth process for association football matches. *The Statistician 47*, 523–538.

Dobson, S. and J. Goddard (2001). *The Economics of Football.* Cambridge: Cambridge University Press.

Dyte, D. and S.R. Clarke (2000). A ratings based Poisson model for World Cup simulation. *Journal of the Operational Research Society 51*, 993–998.

Forrest, D., J. Goddard, and R. Simmons (2005). Odds setters as forecasters. *International Journal of Forecasting 21*, 551–564.

Forrest, D. and R. Simmons (2000a). Forecasting sport: the behaviour and performance of football tipsters. *International Journal of Forecasting 16*, 316–331.

Forrest, D. and R. Simmons (2000b). Making up the results: the work of the Football Pools Panel, 1963-1997. *The Statistican 49*(2), 253–260.

Goddard, J. (2005). Regression models for forecasting goals and match results in association football. *International Journal of Forecasting 21*, 331–340.

Goddard, J. and I. Asimakopoulos (2004). Forecasting football match results and the efficiency of fixed-odds betting. *Journal of Forecasting 23*, 51–66.

Goddard, J. and S. Thomas (2006). The efficiency of the UK fixed-odds betting market for Euro 2004. *International Journal of Sports Finance 1*, 21–32.

Karlis, D. and I. Ntzoufras (2003). Analysis of sports data by using bivariate Poisson models. *The Statistician 52*, 381–393.

Koning, R.H. (2000). Balance in competition in Dutch soccer. *The Statistician 49*(3), 419–431.

Kuypers, T. (2000). Information and efficiency: an empirical study of a fixed odds betting market. *Applied Economics 32*, 1353–1363.

Lee, A. (1999). Modelling rugby league data via bivariate negative binomial regression. *Australian and New Zealand Journal of Statistics 41*, 153–171.

Maher, M.J. (1982). Modelling association football scores. *Statistica Neerlandica 36*, 109–118.

Pope, P.F. and D.A. Peel (1989). Information, prices, and efficiency in a fixed-odds betting market. *Economica 56*, 323–341.

Ridder, G., J.S. Cramer, and P. Hopstaken (1994). Down to ten: Estimating the effect of a red card in soccer. *Journal of the American Statistical Association 89*, 1124–1127.

Rue, H. and O. Salvesen (2000). Prediction and retrospective analysis of soccer matches in a league. *The Statistician 49*, 399–418.

Torgler, B. (2004). The economics of the FIFA football World Cup. *Kyklos 57*, 287–300.

7

Hitting in the pinch

Jim Albert
Bowling Green State University

ABSTRACT

In baseball, there is a general fascination regarding the tendency of players to perform well in clutch situations. We look at the clutch-hitting phenomena by focusing on four fundamental rates, the rate of obtaining a walk, the rate of striking out, the rate of hitting a home run, and the rate of getting an "in-play" ball to fall in for a hit. We fit a random effects model to the rates of players for a particular season and show that rates differ with respect to the proportion of variability that can be attributed to luck or chance binomial variation. The success of a hitter can depend on the game situation, and we show how the four hitting rates can depend on these variables. A random effects model with a bias component is used to represent hitting data in "clutch" and "nonclutch" situations. Using this model, we look for players who perform unusually well in clutch situations.

7.1 INTRODUCTION

In sports, there is a general interest in the ability of athletes to perform well in "clutch" or important situations. In a baseball game, the objective for a team is to score more runs than its opponent. There will be particular times during a game when a batter has good opportunities to produce runs, and it is very desirable for him to do well in these clutch situations. For example, consider a single, a baseball hit where the batter reaches first base safely. The value of a single, from the viewpoint of scoring runs, varies greatly depending on the runner's situation. A single when the bases are loaded is much more valuable to the team than a single with the bases empty. A team would like to have players who are able to hit singles (or other base hits) in these important situations. The ability of a player to "hit in the pinch" has always been important as indicated by Dick Rudolph's comment about batting averages in Lane (1925):

"In my mind, there is nothing more deceptive. I would much rather have on my team a player with an average of 0.270 who was a great hitter when hits were needed than a 0.350 hitter who wasn't much good in a pinch. It's the hits in the pinch that count."

Although baseball players have different levels of performance in clutch situations, most researchers have found little evidence for players to have a higher level

TABLE 7.1
Abbreviations for batting
events in baseball.

Plate appearance	PA
At-bat	AB
Walk or base on balls	BB
Hit by pitch	HBP
Strikeout	SO
Hit	H
Home run	HR

of ability in these important clutch situations. In a typical study of this type, one first defines a measure of clutch performance. For example, one might look at the difference in a player's batting performance in clutch and nonclutch situations

$$\text{DIFF} = \text{Performance in Clutch} - \text{Performance in Nonclutch}.$$

Suppose you collect these differences for a group of players for two consecutive seasons. If you find a significant positive correlation in the clutch performance for two seasons, then one could say that there is some evidence for clutch ability. Unfortunately, most researchers have not found any relationship. In other words, there is little evidence to suggest that clutch performances of baseball players persist across seasons.

In this chapter, we look at the general subject of "hitting in a pitch" at a finer level. Albert (2005) proposed a decomposition of a player's plate appearance that separates out a player's rates in walking, striking out, hitting a home run, and getting an "in-play" ball to fall in for a hit. Section 7.2 through Section 7.4 summarize some recent work about these four hitting rates. One interesting conclusion is that these rates differ with respect to the amount of variation due to difference in abilities relative to the size of the variation due to chance (binomial) variance. In Section 7.5, we explore the differences in these hitting rates relative to the runners on base and the number of outs. In Section 7.6, we fit a random effects model to this rate data for all players in the 2004 baseball season to explore the differences in these rates for scoring and nonscoring situations. In Section 7.7, we use posterior predictive methodology to look for outlying ballplayers who exhibit unusually small or large clutch effects.

7.2   A BREAKDOWN OF A PLATE APPEARANCE: FOUR HITTING RATES

A player comes to bat for a plate appearance. There are a number of different outcomes of this plate appearance with abbreviations displayed in Table 7.1.

First, either the player walks or doesn't walk – his chance of walking is estimated by the walk rate defined by

$$\text{Walk Rate} = \frac{BB + HBP}{PA}.$$

Note that we combine walks and hit by pitches in the formula since each event has the same result of getting the batter to first base without creating an at-bat. Removing walks from the plate appearances, we next record if the batter strikes out or not. We define the strikeout rate as the fraction of strikeouts to the number of at-bats or

$$\text{Strikeout Rate} = \frac{SO}{AB}.$$

With walks and strikeouts removed, we next record if the batter hits a home run or not. The home run rate is defined to be the fraction of home runs for all plate appearances where contact is made by the bat. That is,

$$\text{HR Rate} = \frac{HR}{AB - SO}.$$

With the walks, strikeouts, and home runs removed, we only have plate appearances where the ball is hit in the park. Of these balls put in-play, we record the fraction that fall in for hits – we call this the in-play hit rate or

$$\text{Hit Rate} = \frac{H - HR}{AB - SO - HR}.$$

This hit rate has been described as the Ball-in-play Average by Woolner (2001) and the batting average on Balls in Play (BABIP) by Silver (2004). Much of the interest in this statistic is due to McCracken (2001)'s study that indicated that pitcher's had little control over the outcomes of balls put into play that were not home runs.

Since these rates are defined by sequentially removing walks, strikeouts, and home runs from the plate appearances, they measure distinct qualities of a hitter. Specifically, these rates measure

- the talent to draw a walk;

- the talent to avoid a strikeout;

- the talent to hit a ball out of the park (a home run);

- the talent to hit a ball "where they ain't."

In contrast, traditional hitting statistics confound some of these talents. A batting average confounds three batter talents: the talent not to strikeout, the talent to hit a home run, and the talent to hit an in-play ball for a hit. An on-base percentage confounds the batter's talent to draw a walk with his talent to get an in-play hit and his talent to hit a home run. Since a batter's ability encompasses all of these talents, these four rates may provide useful detailed information about the hitting ability of a player.

7.3   PREDICTING RUNS SCORED BY THE FOUR RATES

Since a team, not an individual, produces runs, one can see the usefulness of these rate statistics in predicting runs by looking at team data. We collect hitting statistics for all 30 teams in the 2005 season and compute the four rate statistics for each team. First, we see how each rate by itself predicts runs scored per game. The individual R^2 values are 0.45 for home run rate, 0.21 for walk rate, 0.17 for hit rate, and 0.00 for strikeout rate. Next, we run a stepwise regression with terms added in order of importance. We first add the home run rate variable with a $R^2 = 0.45$, next we add the hit rate with a total R^2 of 0.58, the strikeout rate with a total R^2 of 0.73, and finally the walk rate with a total R^2 of 0.83. It is interesting that each rate explains a significant portion of the variation in the runs data, even when the other variables are included in the model. The final model is

$$\text{RUNS} = -3.2 + 13.2 \times (\text{BB Rate}) - 12.3 \times (\text{SO Rate}) + 40.9 \times (\text{HR Rate}) + 24.5 \times (\text{H Rate}).$$

How does this model compare with other "bad" and "good" predictors of runs scored? If one uses batting average (AVG) to predict runs scored, then the R^2 value is 0.50, and if we use slugging percentage (SLG), the R^2 value is 0.63. These values are not surprising, since AVG and SLG are relatively poor predictors of runs scored. But if we use the better statistics OPS = OBP + SLG, runs created, and LSLR to predict runs scored, the R^2 values are respectively 0.77, 0.82, and 0.86. (LSLR or least-squares linear regression is the best linear combination of singles, doubles, triples, home runs, and walks in predicting runs.) So our new measure has a similar correlation with runs scored as the good measures OPS, runs created, and LSLR. To see if this is generally true across seasons, we looked at team data for the seasons from 1950 through 2005 and performed this regression study for each season. Also we tried the "five rates," consisting of walk rate, strikeout rate, home run rate, singles rate, and doubles + triples rate to predict runs per game. We found that generally our four rates are superior to both runs created and OPS in predicting runs scored. LSLR generally does better than four rates, but the average improvement is only about 1%. It is surprising to find that "five rates" that distinguishes singles from doubles/triples does marginally better than "four rates" – the average improvement was only 1.4%. This indicates that there is little added value in distinguishing singles from extra-base hits once walks, strikeouts, home runs, and in-play hits are recorded.

7.4   SEPARATING LUCK FROM ABILITY

Suppose we observe a particular rate, say a strikeout rate, for all players in a particular season. For player i, let y_i and n_i denote respectively the number of strikeouts and at-bats. If there are N players, then we will observe much variation in the player strikeout rates $y_1/n_1, ..., y_N/n_N$.

What is the explanation for the variation in the observed rates? One explanation is that the players have different talents to avoid striking out. A second explanation for the spread of strikeout rates is luck or chance binomial variation. We are interested in understanding how much of the total variation in the observed rates is due to differences in the batters *abilities* and how much of the total variation is due to binomial variability.

An attractive way of separating out the ability and luck dimensions of the hitting rates is by use of the following binomial-beta random effects model. We first assume that the counts $y_1, ..., y_N$ are independent, where y_i is distributed binomial with sample size n_i and probability of success p_i. We can think of p_i as representing a player's ability (or lack of ability) of achieving the given event. In the strikeout example, p_i represents the player's ability to avoid striking out. At the second stage of the model, the probabilities $p_1, ..., p_N$ are assumed to be a random sample from a talent distribution $g(p)$. Since p is a proportion, it is convenient to let g be the beta family

$$g(p) = \frac{1}{B(a,b)} p^{a-1}(1-p)^{b-1}, 0 < p < 1,$$

where $B(a, b)$ is the Beta function. It is helpful to reparameterize the beta parameters a and b by the precision $K = a + b$ and the mean $m = a/(a + b)$. To complete the model, we assign vague prior distributions on the prior parameters K and m. We assume K, m are independent with K assigned the proper vague density $g(K) = 1/(K + 1)^2$ and m the improper vague prior $g(m) = m^{-1}(1 - m)^{-1}$.

In this analysis, we only consider players in the 2004 season with at least 100 plate appearances. This excludes pitchers who likely have different hitting characteristics from nonpitchers. For each set of player rates (walk rates, strikeout rates, home run rates, and in-play hit rates), we fit the binomial/beta random effects model.

Table 7.2 gives estimates of the beta parameters a and b that define the talent distribution of all players. The amount of talent inherent in a particular rate can be measured by the estimate of the precision parameter $K = a + b$. The strikeout rate distribution has the smallest estimated value of K which indicates that the value of this statistic is heavily controlled by the ability of the hitter. In contrast, the in-play hit rate has a large estimated value of K that indicates that this statistic is controlled largely by chance variation. One way of quantifying the ability/luck split is by computing the percentage of luck of the players' rates with a given number of opportunities n:

$$\text{PCT LUCK} = 100 \times \frac{\hat{m}(1 - \hat{m})/n}{\hat{m}(1 - \hat{m})/n + \hat{m}(1 - \hat{m})/(\hat{K} + 1)}$$

$$= 100 \times \frac{\hat{K} + 1}{n + \hat{K} + 1}.$$

(In this expression, $\hat{m}$ is the estimated mean value of the beta mean parameter m.) Table 7.3 shows the percentage of luck of each type of batting statistic for different number of opportunities to hit. To help interpret this table, suppose that one has a collection of strikeout rates of hitters all with 250 at-bats. By the table, only 14% of the variability of the rates is due to binomial (chance) variation – the remainder is due to the differences in the abilities of the hitters. In contrast, suppose you have the

TABLE 7.2
Fitted beta ability distributions from 2004 data, players with at least 100 pa's.

	$\hat{a}$	$\hat{b}$	$\hat{K}$	p_5	p_{50}	p_{95}
Walk rate	8.74	81.80	90.5	0.051	0.094	0.152
Strikeout rate	7.50	32.82	40.3	0.096	0.181	0.294
Home run rate	2.88	69.23	72.1	0.011	0.036	0.083
In-play hit rate	164.63	390.61	555.2	0.265	0.296	0.329

TABLE 7.3
Percentage of luck of different rate statistics for many players with 100, 250, 500 opportunities to bat.

	$n=100$	$n = 250$	$n = 500$
Walk rate	48	27	15
Strikeout rate	29	14	8
Home run rate	42	23	13
In-play hit rate	85	69	53

in-play hit rates of batters with 250 balls in-play. By the table, 69% of the variability of these rates is due to chance and only 31% is due to differences in the abilities of the hitters to place balls for hits. As one would expect, the percentage of luck statistic is dependent on the size of the sample; for batters with only 100 balls in-play, most (85%) of the variability in the hit rates is attributed to luck.

The above estimated talent distributions give us some understanding about the variation in talent for a particular batting skill. We can also use the estimated values of a and b to obtain estimates for the batting probabilities for the players. For example, suppose we wish to estimate the probabilities of walking, striking out, hitting a home run, and getting an in-play hit for Johnny Damon for the 2004 season. His 2004 hitting statistics are displayed in Table 7.4. From this model, an estimate at a player's probability is given by

$$\hat{p}_i = \frac{y_i + \hat{a}}{n_i + \hat{a} + \hat{b}} = \frac{n_i}{\hat{K} + n_i}\frac{y_i}{n_i} + \frac{\hat{K}}{\hat{K} + n_i}\hat{m},$$

where $\hat{K} = \hat{a} + \hat{b}$, and $\hat{m} = \hat{a}/(\hat{a} + \hat{b})$. This general form can be applied to each of the four rates. The estimate at Damon's strikeout probability is given by

$$\hat{p}_{so} = \frac{71 + 7.5}{621 + 7.5 + 32.83} = 0.119.$$

Note that this is a modest change from Damon's strikeout rate of $71/621 = 0.114$;

TABLE 7.4
Hitting statistics for Johnny Damon for the 2004 season.

AB	H	HR	BB	HBP	SO
621	189	20	76	2	71

TABLE 7.5
Damon's observed rates and probability estimates for each of the four rates for the 2004 season.

Rate	Observed Rate	Probability Estimate
Walk	0.112	0.110
Strikeout	0.114	0.119
Home run	0.036	0.037
In-play Hit	0.319	0.307

the change is modest since we learned that strikeout rates are an ability statistic and relatively unaffected by chance variation. In contrast, the estimate at Damon's in-play hit rate is given by

$$\hat{p}_{IPH} = \frac{169 + 164.63}{530 + 164.63 + 390.61} = 0.307.$$

Here this estimate is a larger change from Damon's observed in-play hit rate of 0.319; we move his observed rate more towards the average since in-play rates are more controlled by chance. Table 7.5 displays Damon's observed rates and probability estimates for each of the four rates.

7.5 SITUATIONAL BIASES

In the last section, we have gained an overall understanding about the ability dimension of batting rates. Players have different abilities to walk, to strike out, to hit a home run, and to get a batted ball to fall in for a hit, and the distribution of abilities can be described by a talent distribution.

Do these hitting abilities depend on the outs and runners situation? There are three possible outs situations when a player comes to bat (no outs, 1 out, 2 outs) and there are eight possible runner situations since each base can either by occupied or not. So there are $3 \times 8 = 24$ possible inning situations; and it is of interest to say if the four rates depend on the situation.

To gain some general understanding about these situational biases, we initially consider results for all players; and in Section 7.7 we focus on characteristics for individuals. We again consider only the players who had at least 100 plate appearances in the 2004 season. Table 7.6 through Table 7.9 display the overall walk rates,

TABLE 7.6
Walk rates as function of inning situation for players with at least 100 pa's.

Outs	none	1	2	1,2	3	1,3	2,3	full
				Runners on base				
0	0.082	0.069	0.094	0.078	0.112	0.100	0.129	0.075
1	0.084	0.080	0.162	0.090	0.137	0.085	0.263	0.070
2	0.100	0.092	0.193	0.106	0.191	0.123	0.216	0.094

strikeout rates, home run rates, and in-play hit rates for all 24 outs/runner situations. Figure 7.1 through Figure 7.4 graph these rates as a function of the number of outs to make the patterns more obvious. In the figures, the rates for the "no outs, no runners" situation are displayed using a thin solid line. One should be careful about some of the patterns since small differences in the rates are probably nonsignificant due to sampling variation. But there are several large "significant" differences in the rates that are summarized below. The same conclusions were found by a similar study on the 2005 season data.

- One's tendency to **walk** depends on the runners on base. One is more likely to walk when there are runners on and first base is open (runner on 2nd, runner on 3rd, and runners on 2nd and 3rd). When there is a single runner on 2nd or 3rd, then the chance of walking increases with more outs. One is most likely to walk with runners on 2nd and 3rd and one out.

- Generally, the chance of **striking out** increases with the number of outs. It is unlikely to strike out with runners on 1st and 3rd, and one is more likely to strike out with a runner on 2nd or runners on 2nd and 3rd.

- The tendency to **hit a home run** does not generally depend on the runners/outs situation. But there is an interesting pattern with runners on 2nd and 3rd — the home run rate is low for one out and high for two outs. The chance of hitting a home run with no runners on decreases slightly from 0 to 2 outs.

- The **in-play hit** rates seem to vary by the runners/outs situation. With no outs, it is more likely for a batter to get a hit with a runner on 3rd and less likely to hit with a runner on 2nd or runners on 1st and 2nd. The differences in hit rates for different runner situations are smaller with two outs.

The focus in baseball is how a batter performs with runners on base, or runners in scoring position (at least one runner on 2nd or 3rd base). Table 7.10 displays the hitting rates for batters with runners in scoring position or not, and the rates for runners on base or not. We see some effects in this table that parallel the earlier discussion. It is more likely for a batter to walk with runners in scoring position or with runners on base.

TABLE 7.7
Strikeout rates as function of inning situation for players with at least 100 pa's.

Outs	none	1	2	Runners on base 1,2	3	1,3	2,3	full
0	0.168	0.149	0.152	0.150	0.162	0.135	0.167	0.160
1	0.185	0.172	0.187	0.162	0.174	0.153	0.156	0.158
2	0.199	0.179	0.214	0.189	0.218	0.188	0.217	0.194

TABLE 7.8
Home run rates as function of inning situation for players with at least 100 pa's.

Outs	none	1	2	Runners on base 1,2	3	1,3	2,3	full
0	0.043	0.038	0.031	0.034	0.031	0.036	0.033	0.032
1	0.042	0.042	0.038	0.043	0.040	0.035	0.024	0.031
2	0.040	0.040	0.040	0.037	0.036	0.039	0.058	0.046

TABLE 7.9
In-play hit rates as function of inning situation for players with at least 100 pa's.

Outs	none	1	2	Runners on base 1,2	3	1,3	2,3	full
0	0.301	0.303	0.260	0.267	0.358	0.316	0.316	0.311
1	0.298	0.317	0.282	0.299	0.329	0.304	0.301	0.293
2	0.294	0.306	0.284	0.285	0.286	0.298	0.274	0.288

TABLE 7.10
Hitting rates in nonscoring, scoring position, no runners, and runners on.

Rate	Situation Nonscoring	Scoring	No runners	Runners on
Walk	0.086	0.135	0.087	0.113
Strikeout	0.178	0.181	0.181	0.175
Home run	0.042	0.038	0.042	0.039
In-play hit	0.301	0.289	0.298	0.298

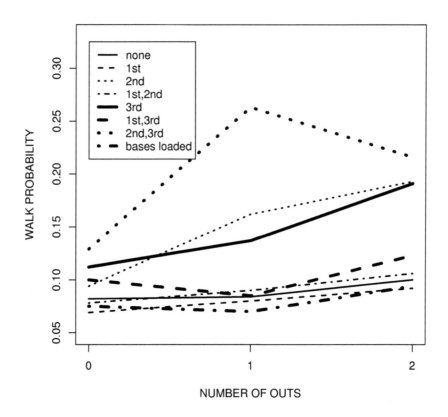

FIGURE 7.1
Walk rates by number of outs and runners on base.

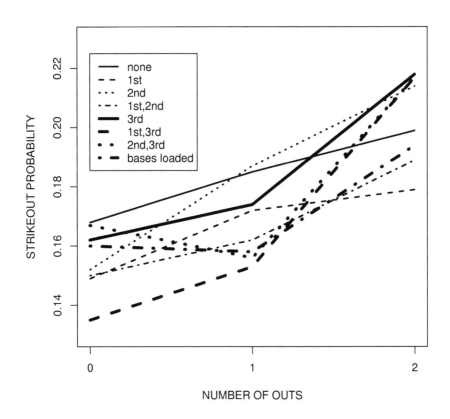

FIGURE 7.2

Strikeout rates by number of outs and runners on base.

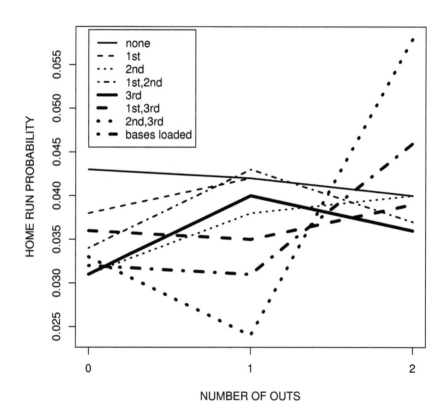

FIGURE 7.3
Home run rates by number of outs and runners on base.

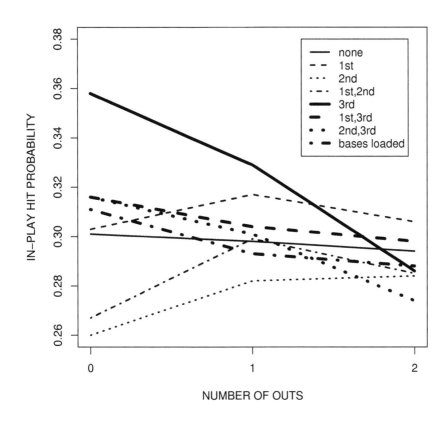

FIGURE 7.4

Hit rates by number of outs and runners on base.

There is a small increase in the strikeout rate with runners in scoring position, and a small decrease in the strikeout rate with runners on base. It is more likely to hit a home run in "nonclutch" situations, and it is less likely to get a in-play hit when runners are in scoring position.

7.6   A MODEL FOR CLUTCH HITTING

We have seen in the above section that there are significant hitting situational effects. The runners on base and the number of outs can affect the tendencies to walk, strike out, homer, and hit. But what does this say about the players' *abilities* to take advantage of the situation? We say that a player is *clutch* if his ability to succeed in a clutch situation, such as runners in scoring position, exceeds his ability to succeed in a nonclutch situation. If the player has a greater ability to succeed in a nonclutch situation, he will be called a *nonclutch* player. In this section, we will regard scoring position and not-scoring position respectively as clutch and nonclutch situations.

Albert and Bennett (2003) gives a general discussion of situational effects. They looked at a number of situations such as home vs away, the pitch count, grass vs turf, and day vs night games. For a given situation, say home and away, one can define the ith player's hitting probability p_i^{HOME} at home and his hitting probability p_i^{AWAY} for away games. There are three possible models for these situational probabilities:

- **No Effect Model.** Here there is no general tendency to hit better at home, or $p_i^{HOME} = p_i^{AWAY}$.

- **Bias Model.** Here there is a general tendency to hit better at home, but the increase is the same for all hitters. That is $p_i^{HOME} = p_i^{AWAY} + b$, where b is the bias.

- **Ability Model.** Here players have different abilities to do better at home. That is, $p_i^{HOME} = p_i^{AWAY} + b_i$, where the individual biases $\{b_i\}$ come from an ability distribution $h(b)$.

Albert and Bennett (2003) fit these models to the number of hits (H) and at-bats (AB) for all players in a single season for each of the situations. Most of the breakdowns such as home vs away and pitcher of the opposite arm/same arm were classified as bias situations. This means that there is a general tendency to hit (for average) at home or against a pitcher of the opposite arm, but the increase in hitting probability is the same for all hitters.

Since hitting success can vary by the bases and outs situation, the following random effects model with a bias component was used to model clutch hitting data. Let y_{i0} and y_{i1} denote the number of successes (out of n_{i0} and n_{i1} opportunities) of the ith player in nonclutch and clutch situations. We assume that y_{ij} is binomial(n_{ij}, p_{ij}), $j = 0, 1$. Let $\eta_{ij} = \log(p_{ij}/(1 - p_{ij}))$ denote the logit of the success probability. The nonclutch and clutch logits for the ith player are given the linear models

$$\eta_{i0} = \theta_i, \ \eta_{i1} = \theta_i + \beta,$$

TABLE 7.11
Estimated parameters and standard errors using logistic random
effects model with a bias component.

	$\hat{\mu}$ (se)	$\hat{\beta}$ (se)	$\hat{\tau}$ (se)
Walk rate	−2.44 (0.02)	0.51 (0.02)	0.35 (0.02)
Strikeout rate	−1.63 (0.02)	0.01 (0.02)	0.42 (0.03)
Home run rate	−3.31 (0.04)	−0.14 (0.03)	0.62 (0.03)
In-play hit rate	−0.85 (0.009)	−0.06 (0.01)	0.09 (0.01)

where $\theta_1, ..., \theta_N$ are a random sample from a normal distribution with mean μ, and
standard deviation τ and β is the bias constant. To complete the model, β, μ, and
$\log \tau$ are assigned flat noninformative priors. Essentially this model says that there
are different ability distributions for the hitting probabilities in the nonclutch and
clutch situations, where the only difference in the distributions is a shift by a constant
value β.

This logistic random effects model was fit to each of the four datasets. The estimated parameters and standard errors are displayed in Table 7.11. The parameter
estimates mirror the patterns that we saw in the hitting rates in Table 7.10.

To help understand the fitted model, consider the estimates for the walk rates.
From this model, the logits of the walk probabilities have a normal(−2.44, 0.035)
ability distribution in nonclutch situations, and a normal(−2.44+0.51, 0.035) ability
distribution in clutch situations. Figure 7.5 displays these two ability distributions on
the probability (p) scale. Players are generally more likely to walk with runners in
scoring position, but this effect is the same for all players.

7.7 CLUTCH STARS?

The above fitted models are helpful in understanding the general effects in being in
a clutch situation. When runners are in scoring position, one is more likely to obtain
a walk, and one is less likely to hit a home run or to get an "in-play" hit. There is no
evidence that batters are more likely to strikeout in a clutch situation.

But do these bias models explain all of the variation in the observed hitting rates?
Specifically, are there particular players that have unusually high or unusually low
differences in rates (between clutch and nonclutch situations) that are not explained
by these bias models? We answer these questions by checking the model by use of
the posterior predictive distribution. This method is based on simulating hitting data
for future sets of players with the same number of hitting opportunities as the 2004
players, and then seeing if the simulated hitting data looks similar to the observed
hitting data. If the simulated hitting data looks similar to the observed data, this says
that the bias model provides a good fit, and there are no "clutch stars."

To do this model checking, we first simulate parameter values (μ, β, τ) from the
fitted posterior distribution and use these values to obtain values of the hitting probabilities $\{p_{ij}\}$. Then we simulate hitting data $\{y_{ij}^*\}$, where y_{ij}^* is binomial with sample

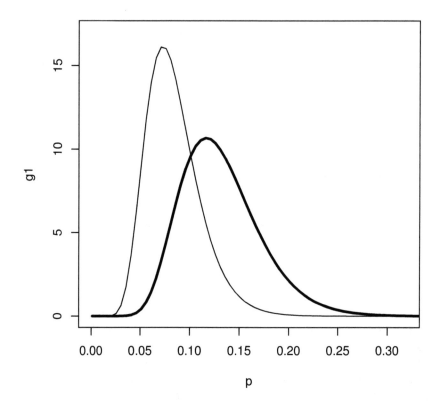

FIGURE 7.5
Fitted ability distributions from bias model for walking probabilities in nonclutch
(light) and clutch (bold) situations.

size n_{ij} and probability of success p_{ij}. From this we can compute the difference in clutch and nonclutch hitting rates

$$d_i^* = \frac{y_{i1}^*}{n_{i1}} - \frac{y_{i0}^*}{n_{i0}}$$

for all players.

The above paragraph describes one simulation of the hitting rates for all players. We repeat this procedure 1000 times, obtaining 1000 sets of the differences in hitting rates $\{d_i^*\}$. To see if these simulated differences resemble the observed differences in rates from the 2004 season

$$d_i^{obs} = \frac{y_{i1}}{n_{i1}} - \frac{y_{i0}}{n_{i0}},$$

we construct a probability plot that graphs the expected order statistics of the $\{d_i^*\}$ (horizontal) against the observed differences in rates $\{d_i^{obs}\}$ (vertical). If the model is a suitable fit, then all of the points should fall on the line through the origin with unit slope. If points fall above the line for large values or below the line for small values, then this indicates that there are some outliers in the data that are not explained by the fitted model. We focus here on the outliers on the high end — these are the players who perform unusually well in clutch situations.

Figure 7.6 displays the probability plot that summarizes the posterior predictive check of the bias model for the walk rates. Here we see a cluster of points above the line for large values and an obvious outlier. This indicates that there are a number of players who have unusually high clutch vs nonclutch walk rates. Table 7.12 displays ten players who have these large differences in rates. Barry Bonds is the obvious outlier — pitchers are afraid of pitching to Bonds with runners in scoring position, and he is much more likely to walk in this situation. To a lesser degree, pitchers seem to be afraid of some other hitters in this list, or perhaps these players have an unusually good talent of working a walk with runners in scoring position.

When one looks at strikeout rates, one is interested in finding outliers at the low end — these correspond to hitters who are less likely to strike out with runners in scoring position. There is a cluster of eight points below the line at the low end of the probability plot of Figure 7.7, and Table 7.13 displays these "clutch" hitters.

Generally the bias models seem to provide a better fit for the two remaining rates. Figure 7.8 displays the probability plot for the home run rates. We are interested in looking for outliers at the high end – there are only two points above the line in the graph. As Table 7.8 indicates, Mark Teixeira and Jacque Jones are the only two players who seem to be unusually more effective in hitting home runs in the clutch situation. When one looks at the probability plot for in-play hit rates (Figure 7.9), there are no points above the line at the high end. There is little evidence that any player can do unusually better in getting in-play hits with runners in scoring position.

7.8 RELATED WORK AND CONCLUDING COMMENTS

As mentioned in the introduction, there have been many studies that have found little evidence of clutch hitting ability in baseball. Examples of these studies include

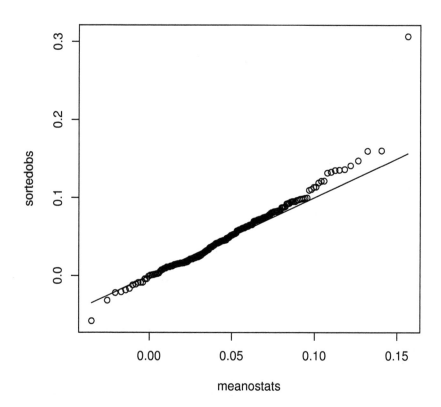

FIGURE 7.6
Probability plot for the difference in hitting rates for the walk rates.

TABLE 7.12

Unusually high clutch hitters with respect to walk rate for 2004 season.

Players	clutch rate − nonclutch rate
Barry Bonds	0.307
Luis Gonzalez	0.160
Mike Piazza	0.160
Ichiro Suzuki	0.148
Todd Helton	0.141
Ray Durham	0.136
Khalil Greene	0.135
David Bell	0.135
Brad Ausmus	0.133
Adam Dunn	0.132

TABLE 7.13

Unusually high clutch hitters with respect to strikeout rate for 2004 season.

Players	clutch rate − nonclutch rate
Richard Hidalgo	−0.104
Chad Tracy	−0.103
Mark Bellhorn	−0.092
Carlos Guillen	−0.084
Matt Holiday	−0.081
Bobby Abreu	−0.080
Bobby Higginson	−0.077
Torii Hunter	−0.076

TABLE 7.14

Unusually high clutch hitters with respect to home run rate for 2004 season.

Players	clutch rate − nonclutch rate
Mark Teixeira	0.077
Jacque Jones	0.051

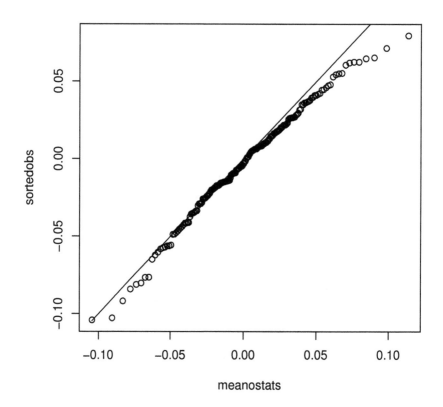

FIGURE 7.7
Probability plot for the difference in hitting rates for the strikeout rates.

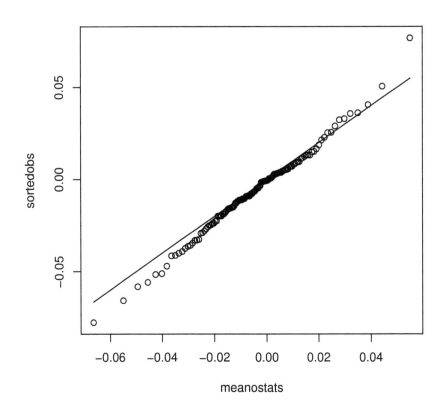

FIGURE 7.8
Probability plot for the difference in hitting rates for the home run rates.

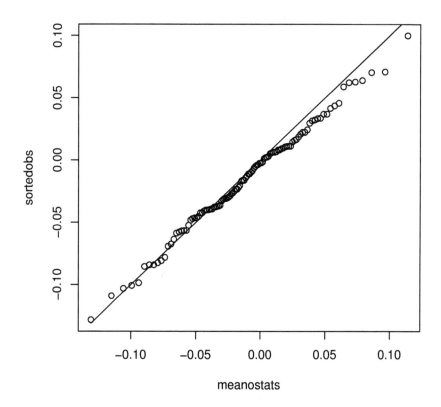

FIGURE 7.9
Probability plot for the difference in hitting rates for the in-play hit rates.

Cramer (1977), Grabiner (1993), and Ruane (2005). Albert (2002) used 1987 National League data to look for the existence of clutch hitting ability. He first defined the value of a plate appearance as the difference in run potentials before and after the plate appearance. For all players, he computed the average value of a plate appearance for all situations with different runners on base and number of outs. He fit a random effects model (similar to the one used in this paper). In fitting this model, he found some situations where batters tend to be more or less successful. But he found little evidence that players had an ability to perform better in particular situations.

James (1984) wondered if it was meaningful to search for clutch hitting ability since the meaning of this hitting characteristic is not clear. One intent of this paper was to try to add some insight into clutch hitting by looking at hitting at a finer level. It is clear from our empirical work that it is more likely to walk with runners on base, and a player's tendency to strike out, to hit a home run, and to get an in-play hit can depend on the bases and outs situation. Generally, random effects models with a bias component seem to provide reasonably good fits to walk, strikeout, home run, and hitting data for all players. But we did find a few outlying players who performed at a higher level than predicted from our models — these outliers primarily occurred in walk and strikeout rates. Some of these outliers are easily understood; for example, Barry Bonds is very likely to walk with runners in scoring position since he is a feared hitter. But some of these other "clutch stars" are not as well understood.

In a recent analysis, Silver (2006) revisits this clutch ability issue with a more sophisticated analysis. He develops a regression model for hitters that combines the number of runs he produces, the number of wins he creates for his team, and his opportunities to get hits in clutch situations. He develops a statistic called "clutch" that measures the results of a player's clutch hitting over and above what would be predicted by his batting statistics, opportunities, and run-scoring environment. He makes the interesting conclusion that 10% of clutch-hitting performance can be explained by skill.

There appears to be a connection of our results with the clutch hitting conclusions of Silver (2006). Silver makes the conclusion that clutch hitting does exist, although its size is relatively small, similar to the size of baserunning ability. He finds that there is a significant relationship between his clutch rating and walk and strikeout rates — batters who walk often and strike out rarely appear to do well in the clutch. This conclusion seems similar in spirit with our analysis in which we found clutch stars with respect to these two batting rates.

REFERENCES

Albert, J. (2002). Hitting with runners in scoring position. *Chance 15*, 8–16.

Albert, J. (2005). Does a baseball hitter's batting average measure ability or luck? *Stats 44*.

Albert, J. and J. Bennett (2003). *Curve Ball: Baseball, Statistics, and the Role of Chance in the Game* (revised ed.). New York: Springer-Verlag.

Cramer, R.D. (1977). Do clutch hitters exist? *Baseball Research Journal 2*.

Grabiner (1993). Clutch hitting study. `http://www.baseball1.com/bb-data/grabiner/fullclutch.html`.

James, B. (1984). *The Bill James Baseball Abstract*. New York: Ballantine Books.

Lane, F.C. (1925). *Batting*. Cleveland, Ohio: Society for American Baseball Research.

McCracken, V. (2001). Pitching and defense: how much control do hurlers have? Baseball Prospectus. `http://www.baseballprospectus.com`.

Ruane, T. (2005). In search of clutch hitting. `http://retrosheet.org/Research/RuaneT/clutch_art.htm`.

Silver, N. (2004). Lies, damned lies: The unique Ichiro. Baseball Prospectus. `http://www.baseballprospectus.com`.

Silver, N. (2006). Is David Ortiz a clutch hitter? In J. Keri (Ed.), *Baseball Between the Numbers*, Chapter 1-2, pp. 14–34. New York: Basic Books.

Woolner, K. (2001). Counterpoint: Pitching and defense. Baseball Prospectus. `http://www.baseballprospectus.com`.

8

Does momentum exist in a baseball game?

Rebecca J. Sela
New York University

Jeffrey S. Simonoff
New York University

ABSTRACT

The nature of baseball is that it is characterized by intermittent periods of action, separated by periods of inactivity. To a fan of the game, it often appears that this results in a sense of momentum, in the sense that the results of previous plays seem to have a relationship to ones that follow. The existence of such an effect can be explored using play-by-play data. In this paper we use multinomial logistic and negative binomial regression models to explore the question of whether momentum effects appear in major league baseball games. We find that the evidence for such effects is weak at best, implying little chance of making meaningful improvements in the prediction of future events using models that include such effects. We also explore the questions of whether leadoff walks are rally starters and double plays are rally killers, finding no evidence to support the former notion, but finding in the latter case evidence of meaningful gains in accuracy when incorporating the knowledge that a double play has occurred into predictive models for run scoring.

8.1 INTRODUCTION

In contrast to many other sports, baseball is a game of intermittent action, separated by (sometimes extended) periods of time between pitches, where (usually) nothing happens. While for some this leads to boredom, for baseball fans these times of inaction increase anticipation, particularly at times when one team is building a scoring rally through several at bats.

The natural impression a fan often gets as a team puts together several hits or walks and scores some runs is one of momentum — a sense of the inevitability of yet another hit leading to still more runs — but is that really the case? Does the outcome of a particular at bat depend on what has come before, or is it just a function of the current out and runners on base situation (the current *state*), current pitcher, batter, and so on? This question can be explored using play-by-play data, starting with simple models for the transition from one state to another that depend only on the initial state, then extending those models to include other characteristics of the current situa-

135

ation (situational effects), and finally including effects depending on previous at bats (momentum effects). This is different from questions of whether individual players or teams are streaky (that is, they have periods of time when they are more or less successful than would be expected by chance), which have been examined, for example, by Albright (1993) and in Chapter 9 of Albert and Bennett (2003), but is related to the same issue of whether fans' perceptions of non-randomness are real or not.

In the next section we describe how batting transitions from one batter to the next can be modeled statistically. We outline the simplest model for transitions from one state to another, the Markov chain, and then discuss how this model can be generalized to allow for situational and momentum effects. The models are then fit to actual play-by-play data, in order to see if there is evidence for situational and momentum effects. We also examine models for scoring that allow for momentum effects, and explore whether certain plays can be viewed as rally starters or rally killers (and hence momentum-causing events).

8.2   MODELS FOR BASEBALL PLAY

A simple model that has been used in the past to represent baseball play (apparently starting with Howard, 1960) is the Markov chain. Each half-inning of a baseball game starts with no outs and no runners on base. As the game progresses, the state changes, depending on the result of each at bat. Since there are three possible numbers of outs (0, 1, or 2) and eight possible runner locations (three bases, each of which might be occupied or unoccupied), there are $3 \times 8 = 24$ possible states, plus the final absorbing state of 3 outs. Thus, for any situation in a game, the possible transition to the next situation (after the batter bats, for example) can be represented by a 24×25 probability matrix $P = (p_{ij})$, where p_{ij} is the probability of going from state i to state j. Of course $\sum_j p_{ij} = 1$; note also that not all of the probabilities are positive, since (for example) it is not possible to go in one step from the {no outs, no runners on base} state to any states with two outs. In fact, only 308 of the 600 possible transitions have positive probabilities, and many others are extremely unlikely. A simpler version of this representation is to record only the number of runners on base, rather than the bases occupied (resulting in only $3 \times 4 = 12$ initial states, and positive probabilities for 74 of the 156 possible transitions), since the importance of a plate appearance appears to be much more related to the number of baserunners than to which bases they occupy (Albert, 2001).

The simplest model for baseball transitions is based on assuming that the transition to a new state depends only on the current state, resulting in a Markov chain (see Chapter 12 of Albert, 2003, for a detailed description of the application of this model). Although this model provides a reasonable first representation of the data, it clearly cannot be a complete representation of transitions, since it does not take into account important factors such as the quality of the pitcher, the quality of the hitter, how long the pitcher has been in the game, and so on. Such factors can be incorporated into the model through the use of a series of multinomial logistic regression models (Simonoff, 2003, Chapter 10). Consider a particular starting state i, and let $p_{ij}^*, j = 1, \ldots, J$ be the probability of transitioning to the jth state, assuming that probability is positive (more correctly J depends on i, but we will suppress that for ease of notation). For any game situation starting in state i, the response y is the state category to which there is a transition. The goal is to construct a model for p_{ij}^* as a function of a set of predictor values $\mathbf{x}$. The model is based on a set of logistic regressions, where the logit is defined relative to one of the transition states, which is termed the baseline category. The choice is ultimately irrelevant from the point of view of model fitting, but interpretation of regression coefficients is easier if one of the categories can be viewed as a "standard;" in this context, it seems natural to use the state that corresponds to the same runner locations but one additional out, since that is always a possible transition and corresponds to a completely unsuccessful batting outcome (an "unproductive out").

Say the baseline category is the J th category. The logistic regression model is

$$\log\left(\frac{p_{ij}^*}{p_{iJ}^*}\right) = \beta_{0j} + \beta_{1j}x_1 + \cdots + \beta_{kj}x_k, \qquad j = 1, \ldots, J - 1. \qquad (8.1)$$

The model incorporates $J - 1$ separate equations, each of which is based on a distinct set of parameters $\boldsymbol{\beta}$. For the baseline category J, $\beta_{0J} = \beta_{1J} = \cdots = \beta_{kJ} = 0$. Model (8.1) implies a simple functional form for these probabilities. The form is the familiar S-shape for a logistic relationship,

$$p_{ij}^* = \frac{\exp(\beta_{0j} + \beta_{1j}x_1 + \cdots + \beta_{kj}x_k)}{\sum_{\ell=1}^{J} \exp(\beta_{0\ell} + \beta_{1\ell}x_1 + \cdots + \beta_{k\ell}x_k)}. \qquad (8.2)$$

The model is fit to play-by-play data by fitting separate models for all events from each of the initial states i. Note that the simple Markov chain model includes no predictors $\mathbf{x}$; in that case, the fitted probabilities p_{ij}^* are just the empirical proportions of times state i transitioned to state j in the data, as would be expected. Koop (2004) also discussed how the Markov chain model can be generalized to multinomial regression models to account for team-related effects in baseball data, although in his case the transitions were from year-to-year, with the states having a natural ordering. The estimates $\{\boldsymbol{\beta}_1, \ldots, \boldsymbol{\beta}_{J-1}\}$ are obtained using maximum likelihood, where the log-likelihood is

$$L = \sum_{j=1}^{J} \sum_{y_\ell=j} \log p_{ij(\ell)}^*,$$

where the second summation is over all observations ℓ with response level j, and $p_{ij(\ell)}^*$ is the probability (8.2), substituting in the predictor values for the ℓth observation.

Since the data sets being used here are large (involving several hundred thousand transition events), formal hypothesis tests can be misleading when attempting to choose among different models. For this reason, two other approaches that are less sensitive to the availability of very large samples will be used. The first approach is through the use of statistical information (Burnham and Anderson, 2002; Linhart and Zucchini, 1986; McQuarrie and Tsai, 1998). The Akaike Information Criterion

(*AIC*) explicitly trades off goodness-of-fit of a model with parsimony by supporting models with smaller values of

$$AIC = -2L + 2\gamma,$$

where L is the log-likelihood function, and γ is the number of estimated parameters in the model. Selecting a model to minimize *AIC* achieves the goal of balancing goodness-of-fit (through higher L) with parsimony (through smaller γ). Models with smaller values of *AIC* are preferred over those with larger values, although in general any models with values of *AIC* within 3–5 of each other can be viewed as indistinguishable from a practical point of view.

The second approach to evaluating the worth of models is that of validation, where models fit on a training set are then compared based on their performance on a new test set. In the present context, this is naturally accomplished by building the models based on one set of years, and then validating it on the following year's data.

In addition to these multinomial logistic regression models, count regression models (Simonoff, 2003, Chapter 5) are also used here to investigate momentum. In these models the relationships between predictors and the number of runs scored from a given play onwards are explored, in order to track which variables (given the initial state) are related to the ability to score runs. The most commonly used count regression model is based on the Poisson random variable, where the number of runs scored y_i for the ith event is modeled as a Poisson random variable with mean μ_i, and the logarithm of μ_i is linearly related to a set of predictors, implying

$$y_i \sim \text{Poisson}[\exp(\beta_0 + \beta_1 x_{1i} + \cdots + \beta_p x_{pi})].$$

The Poisson model assumes that all heterogeneity in expected runs scored is reflected in information from the predictors that are in the model, but this is often untrue. As a result of this unmodeled heterogeneity, overdispersion occurs, where the observed variance in counts is larger than would be implied by the mean, and alternative models are required. The most common parametric model for overdispersion is based on the negative binomial distribution, with probability distribution

$$f(y_i; \mu_i, v) = \frac{\Gamma(y_i + v)}{y_i! \Gamma(v)} \left(\frac{v}{v + \mu_i} \right)^v \left(\frac{\mu_i}{v + \mu_i} \right)^{y_i},$$

with $\mu = \exp(X\beta)$ and v a parameter that incorporates overdispersion (the negative binomial probability function approaches that of the Poisson as $v \to \infty$). The necessity of going from the Poisson family to the negative binomial family can be tested using easily constructed score tests (Simonoff, 2003, page 148), and *AIC* also can be used to assess the benefit of negative binomial regression models over Poisson models (as well as to help choose among candidate models).

8.3   SITUATIONAL AND MOMENTUM EFFECTS

The models of the previous section can be used to explore the effects of situational and momentum variables using play-by-play data. Such data is available from Retrosheet (www.retrosheet.org), which gives box scores for every major league game from 1957 through 2005. Retrosheet also includes statistics for individual players, giving data for all nonpitchers with at least 75 plate appearances and all pitchers with 25 or more innings pitched in the season.

The modeling approach is to start with the Markov chain model as a baseline, and then see if adding variables provides improved predicting power. The Markov chain model is based only on the current state (the current outs/runners combination). The first generalization is to include situation effects. These effects refer to those that are based on the current situation, but not the current state, the distinction being that the current situation includes information about the pitcher, batter, and game situation. The situational variables included are as follows:

1. The on-base percentage (OBP) and slugging percentage (SLG) of the batter for that season. These two variables have been found in many studies to be very effective measures of the quality of a hitter, particularly studies based on the principles of *sabermetrics*, the statistical analysis of baseball (see the web site for the Society for American Baseball Research, www.sabr.org).

2. The average number of strikeouts per 9 innings (K9IP) and average number of walks-plus-hits per inning (WHIP) of the pitcher for that season. These are prominent sabermetric measures for pitching quality.

3. Whether the team at bat is the home or visiting team.

4. Measures of the fatigue of the pitcher, including the number of pitches thrown and number of batters faced.

5. The OBP and SLG of the batter who bats next. The idea here is to allow for the possibility that certain batters provide protection (or lack of protection) for the batters batting before them because of their own abilities and pitcher's willingness or unwillingness to pitch to them. We use the statistics for the batter who actually batted, rather than those for the batter scheduled to bat, based on the idea that a pitcher who believes that the following hitter will be pinch hit for, will pitch in a way consistent with the abilities of the pinch hitter, rather than the scheduled hitter. If a batter is the final batter of the game, the values used are those of the batter that hit eight batters earlier. If the values for the next batter are missing, they are treated as zero (the idea being that such a player would not be viewed as threatening by the pitcher).

Since there is ample evidence of different hitting and pitching quality between different players, it would be expected that a model including situational effects would be an improvement on the simple Markov model.

In contrast to situational effects, momentum effects address the possibility of predictive power based on information from previous situations. Momentum effects are explored using the following variables:

1. The result of the previous situation (e.g., generic out, strikeout, walk, single, etc.). Since we are interested in momentum, which is unlikely to carry over

from a previous inning's at bats, the previous result for the leadoff hitter of an inning is coded as an "unknown event."

2. The number of batters since the last out.

3. The number of runs scored since the last out.

The latter two variables are intended to capture numerically the sense noted in the introduction of the increasing "inevitability" of a hit when several batters in a row have gotten on base. Momentum models are ones that imply that the current status of the game (including the relative qualities of the pitcher and hitters and the fatigue of the pitcher) is not sufficient to account for the result of the play, and that results of previous plays also matter.

The models are fit to data from the (combined) 2003 and 2004 major league seasons, and are validated on data from the 2005 season. Data from the American League and the National League are treated separately, for two reasons. First, given the difference in rules of play in the two leagues (in the American League the pitcher, who is typically a poor hitter, is replaced in the lineup with a designated hitter, while this is not done in the National League), it is plausible that there could be (hopefully minor) differences in situational and momentum effects in the two leagues. Second, analyzing the data separately provides an additional validation of the models and their implications, since one would expect that any real effects would appear in both leagues. Only events that correspond to batter changes are considered (that is, transitions corresponding to stolen bases, for example, are not included), resulting in roughly 175000 transitions for the 2003-2004 American League data, validated on roughly 90000 transitions in 2005, and roughly 200000 transitions for the 2003-2004 National League data, validated on roughly 100000 transitions in 2005 (the difference is because the National League consists of 16 teams, while the American League includes 14 teams).

8.4 DOES MOMENTUM EXIST?

8.4.1 MODELING TRANSITION PROBABILITIES

We first examine effects related to transition probabilities using multinomial logistic regression models. As noted earlier, these analyses could be based on either the full 24×25 set of transitions, or the simpler 12×13 set. We will focus here on the simpler set of transitions, since they are much easier to understand, and all of the effects described here also appear in the more complex transition models.

Table 8.1 gives *AIC* values for the American League data. Since it is only differences of *AIC* values between models that matter (rather than the actual values, which include various constants that are the same for all models), the numbers given are the differences of *AIC* values from that of the model with minimum *AIC* value. Models are fit to all events consistent with each initial state (number of outs and number of runners on base), and correspond to the Markov model (no predictors), a "situation effects" model based on pitching and hitting statistics for the current pitcher and batter and the home team indicator, a second "extended" situation effects model that

TABLE 8.1

AIC values for American League transition multinomial logistic regression models, and the proportion of 2005 events where the momentum model based on the result of the previous play predicts better than the extended situation effects model.

Outs	Men on base	Markov	Situation effects Basic	Situation effects Extended	Situation effects Full	Momentum effects Previous result	Proportion better
0	0	594	16	**3**	9	6	
0	1	1307	89	97	9	**0**	0.477
0	2	191	186	206	24	**0**	0.506
0	3	**0**	16	57	106	86	
1	0	400	9	**0**	37	32	
1	1	1371	1045	1031	8	**0**	0.449
1	2	502	335	304	5	**0**	0.582
1	3	3	**0**	23	88	73	
2	0	261	3	**0**	29	26	
2	1	712	317	280	4	**0**	0.457
2	2	351	128	96	**2**	8	0.449
2	3	14	**0**	13	48	42	

adds fatigue variables and the hitting quality of the batter up next, and "momentum effects" models. Each of the seven possible momentum models (based on the different possible subsets of momentum predictors added to the extended situation effects model) was fit, but results are only given for the model using all three momentum predictors and the one using only the categorical predictor corresponding to the result of the previous event. This latter model was almost always the best momentum choice in terms of *AIC*, but occasionally it (barely) was not; in such situations the corresponding entry in Table 8.1 for the model could be nonzero, while still being the smallest number for that row in the table. In order to make the *AIC* values comparable, all of the models were fit on the subset of the data that was complete for all of the potential predictors. The *AIC* value for the model with smallest value in the table (and hence the preferred choice) for that initial state is given in bold face.

Several patterns are evident from the table. First, as would be expected, adding situation effects variables almost always improves on the Markov model; that is, the abilities of the pitcher and batter have an effect on the outcome of the current play. Further, while momentum effects apparently are present when there is either one man on base or two men on base (no matter how many outs there are), they are absent when the bases are empty or full (again, no matter how many outs there are). The lack of any momentum effect when the bases are empty implies what could be termed a "reset" effect, in that once the bases are empty (even if it is after several runs have scored) everything starts over as if no runs have scored. The lack of an effect when the bases are loaded suggests that since this is the highest pressure situation for both the pitcher and the batter, it does not really matter if the bases just became loaded with the given number of outs, or if the pitcher has gotten an out or two.

As was noted earlier, the momentum model that is generally strongest is the one that depends on only the previous result. Thus, it appears that momentum (when it exists) is a short-term effect, depending only on what just happened, rather than an accumulation of several successful or unsuccessful at bats. As would be expected, transitions corresponding to successful at bats are more likely to follow successful ones, and transitions that correspond to unsuccessful at bats are more likely to follow unsuccessful ones.

Unfortunately, there is probably less here than meets the eye. While the presence of momentum effects when there are one or two baserunners seems real, that does not mean that it translates into better predictions in new situations. In order to investigate this, the models built on the 2003-2004 data were validated on the 2005 data, in the following way. First, the models were applied to the 2005 data, resulting in estimated probabilities for all of the possible transitions from the current state. Then, for each 2005 event, the actual transition was determined (this was not used in the fitting based on the 2003-2004 data, of course), and the estimated probability for each model for that transition was recorded. The model that has higher estimated probability for the transition that actually occurred is considered the more successful model for that event. The last column of Table 8.1 gives the proportions of the time that the momentum effects model was more successful than the situation effects model for the six circumstances where the momentum effects model fit the 2003-2004 data better. Values greater than 0.5 correspond to situations where the addition of the momentum effect (the result of the previous play) results in improved predictive performance. As can be seen, the only situation where there is a meaningful benefit is when there is one out and there are two baserunners; when there are no outs and two baserunners the two models perform roughly equivalently; and in all other cases the momentum effects model does noticeably worse. Thus, the additional complexity of the momentum effects model, even though it provides real benefits when fitting the model, results in poorer predictions on new data in most situations.

A closer examination of the one favorable situation (one out and two runners on base) shows that the only transitions for which the momentum effects model is better than the situation effects model correspond to unproductive outs — either a transition to two outs and two baserunners, or a transition to three outs and the end of the inning. These are strongly associated with the previous result also being an out. Thus, the only momentum effect that seems to be able to be predicted is the momentum of effective pitching and poor hitting (outs following outs), rather than effective hitting and poor pitching (hits and walks following hits and walks).

Table 8.2 summarizes the results for the National League data. It is encouraging to note that the model fitting to the National League data follows virtually the same pattern as it did to the American League data. That is, the situation effects models almost always improve on the Markov model, but the momentum effects models only improve on the situation effects models when there are either one or two baserunners. Further, the performance when validating on the 2005 data is better than that for the American League data, as the momentum effects model has better predictive

TABLE 8.2
AIC values for National League transition multinomial logistic regression models, and the proportion of 2005 events where the momentum model based on the result of the previous play predicts better than the extended situation effects model.

Outs	Men on base	Markov	Situation effects Basic	Situation effects Extended	Momentum effects Full	Momentum effects Previous result	Proportion better
0	0	912	14	**0**	19	13	
0	1	1408	1077	1004	**1**	1	0.471
0	2	211	161	179	15	**0**	0.437
0	3	**0**	5	25	131	110	
1	0	552	**0**	7	49	42	
1	1	1455	1061	962	7	**0**	0.434
1	2	621	469	425	**5**	7	0.640
1	3	15	**0**	22	85	65	
2	0	394	10	**0**	28	23	
2	1	830	459	90	3	**2**	0.612
2	2	336	183	64	7	**0**	0.549
2	3	15	3	**0**	41	32	

performance when there is one out and two baserunners (as is true for the American League data) and also when there are two outs and either one or two baserunners. Once again, this comes from poor hitting following poor hitting, rather than good hitting following good hitting, as roughly 80%, 79%, and 68%, respectively, of the transitions corresponding to outs from the {one out, two baserunners}, {two outs, one baserunner}, and {two outs, two baserunners} situations, respectively, are better predicted by the momentum effects model.

An interesting question is why the models are more effective for National League data than for American League data. Two explanations seem plausible. First, the presence of the designated hitter means that American League lineups are stronger than those of the National League, especially towards the bottom of the order. Indeed, some American League teams will put a strong hitter last in the lineup, in order for him to be a "second leadoff hitter" (since the ninth spot in the order comes to bat right before the top of the order in any situation other than the first at bat of the game). This could have the effect of reducing a pitcher's momentum, since a relatively weak eighth-place hitter could be followed by a stronger ninth-place one. In contrast, in the National League, the pitcher bats ninth, and he is typically a very poor hitter; thus, if the lineup gets progressively weaker from the seventh to eighth to ninth spots, increased predictive ability based on the previous result being an out becomes more likely. It should be recalled, however, that the quality of the batter is part of the situational (and momentum) effects models, so it is not merely the poor hitting ability of pitchers that is causing an apparent (but not real) momentum effect.

Statistical Thinking in Sports

A second possible reason for the improved ability to predict unsuccessful at bats in the National League is that over the years in question the American League has been viewed as clearly superior to the National League, particularly with respect to hitting. That is, the actual quality of the team as a whole could have some relationship to the existence of momentum.

8.4.2 MODELING RUNS SCORED

In this section we explore whether situational and momentum effects are predictive for the number of runs scored in the inning from the current play onwards. While this is clearly related to the earlier analyses on transitions, it is not the same, as ultimately it is only runs scored that matter to the teams (while the transitions could be viewed as the building blocks of scoring runs). We again focus on the simpler (twelve initial states) data. Initial fitting was based on Poisson regression models, but both score tests and *AIC* strongly identify overdispersion effects, so all of the discussion here is based on negative binomial regression models. The two situation effects models and the full set of momentum effects models were compared in the same way as was done in the previous section. It turns out that the *AIC* values (not reported here) follow precisely the same pattern as was true for the multinomial logistic regression models; that is, the momentum effects models are chosen when there are either one or two baserunners, are not chosen when there are either zero or three baserunners, and the momentum model based on only the result of the previous play is almost always the momentum effects model of choice.

Figure 8.1 and Figure 8.2 (American League data) and Figure 8.3 and Figure 8.4 (National League data) summarize the predictive ability of the models. Each of the figures is based on a comparison of the absolute error in estimated runs scored for the extended situation effects model versus the momentum effects model based only on the result of the previous play. Figure 8.1 and Figure 8.3 plot the proportion of the time the momentum effects model had smaller absolute error, separated by the actual number of runs that were scored from the current play onwards, while Figures 8.2 and Figure 8.4 plot the median difference in absolute errors of the two models (with positive values corresponding to higher absolute errors for the situation effects model). In each plot, events starting with zero outs are given as the solid line, events starting with one out are given by the dashed line, and events starting with two outs are given by the dotted line. Values are only plotted when the proportions and medians are based on at least 15 events (this accounts for the different values of maximum number of runs scored plotted for the different initial states).

For the American League data, when there are no outs, the only times that the momentum effects models are accurate more often than the extended situation effects model (Figure 8.1) is when the inning is relatively low scoring (no further runs if one man is on base, and no more than one further run when two men are on base). This is consistent with the pattern observed earlier, in that the model is correctly predicting fewer runs scored after an out (that is, the momentum reflects outs following outs, rather than hits following hits).

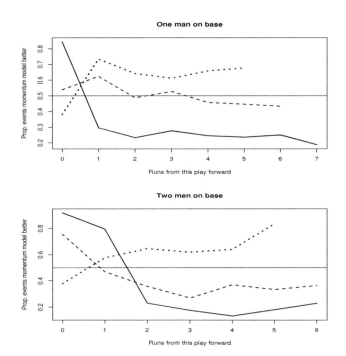

FIGURE 8.1

Proportion of the time that the momentum effects model based on the result of the previous play has smaller absolute error in predicting runs scored than does the extended situation effects model for the 2005 American League data. The solid line corresponds to starting with zero outs, the dashed line corresponds to starting with one out, and the dotted line corresponds to starting with two outs.

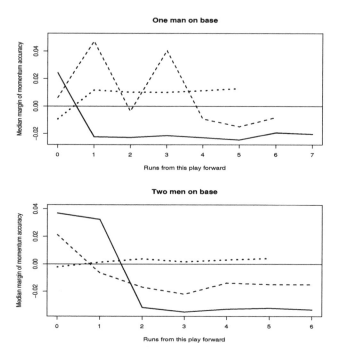

FIGURE 8.2
Median difference in absolute errors of the momentum effects model based on the result of the previous play and the extended situation effects model for the 2005 American League data. Positive values correspond to higher absolute errors for the situation effects model. The solid line corresponds to starting with zero outs, the dashed line corresponds to starting with one out, and the dotted line corresponds to starting with two outs.

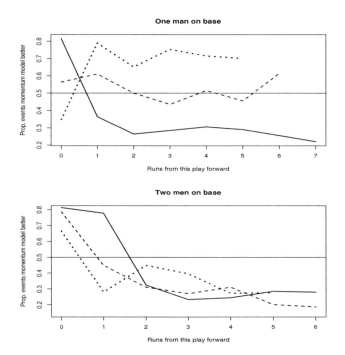

FIGURE 8.3

Proportion of the time that the momentum effects model based on the result of the previous play has smaller absolute error in predicting runs scored than does the extended situation effects model for the 2005 National League data. The solid line corresponds to starting with zero outs, the dashed line corresponds to starting with one out, and the dotted line corresponds to starting with two outs.

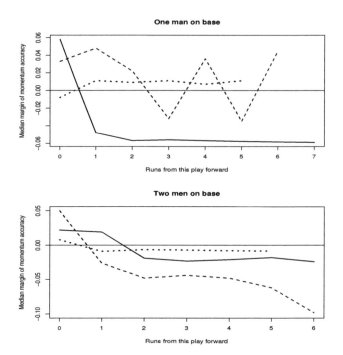

FIGURE 8.4

Median difference in absolute errors of the momentum effects model based on the
result of the previous play and the extended situation effects model for the 2005
National League data. Positive values correspond to higher absolute errors for the
situation effects model. The solid line corresponds to starting with zero outs, the
dashed line corresponds to starting with one out, and the dotted line corresponds to
starting with two outs.

The pattern is similar (although a bit weaker) when there is one out, but when there are two outs, the opposite pattern occurs, and the momentum effects model is more effective for higher scoring innings.

Once again, however, there is less here than meets the eye. Figure 8.2 shows that the gains in predictive power are very small — on the order of 0.02-0.04 runs, including when the actual number of runs scored is quite high. Since the median absolute error of the extended situation effects model is roughly one run when there are no outs, and roughly one-third of a run when there is one out, the gains of the momentum effects model over the situation effects model are also small in a relative sense.

Figure 8.3 and Figure 8.4 summarize the results for the National League data. Once again, the results are broadly similar to those in the American League. When there is one out, the momentum effects model is only an improvement when relatively few runs are scored (no runs if there is one baserunner and zero or one runner when there are two baserunners). The pattern is similar when there is one out, although the effect seems to be weaker than in the American League. When there are two outs the momentum effects model is more effective more often when more runs are scored (Figure 8.3), but the gains in predictive accuracy are small (Figure 8.4). Indeed, while these gains are larger for the National League data than for the American League data, they are still relatively small. Thus, there is little evidence that meaningful gains in predicting runs scored can be achieved by taking into account momentum effects.

8.5 RALLY STARTERS AND RALLY KILLERS

Given the evidence for the existence of momentum in games (weak as it is), a natural question is to wonder what starts and stops such momentum; that is, are certain events rally starters or rally killers? We need to be clear what exactly we mean by this. For example, it is obvious that a leadoff triple is a rally starter, in the sense it results in a transition to a state (no one out with a man on third) that has a higher expected number of runs scored. That is not what we mean in the momentum context, however, since the favorable nature of this for the team at bat does not come from the play itself, but rather the final state (which is the initial state for the next batter).

Rather, in the momentum context, a rally starter is an event that results in higher average runs scored than other events that lead to the same final state. For example, commentators often state that a leadoff walk is a particularly bad thing for a pitcher (the implication being that it is worse than a leadoff single, even though both result in a man on first with no one out), because when a pitcher walks a batter he does not give his fielders a chance to make a play and put the batter out. If this is actually the case, the expected number of runs scored after a leadoff walk should be higher than the expected number of runs scored after a leadoff single (perhaps taking the characteristics of the pitcher and batter into account). Similarly, the double play is often considered a rally killer. At a basic level it is, in that going from no outs to two outs greatly reduces the chances of scoring, but in the context of momentum the question is whether a double play as the previous outcome is worse than an outcome that leads to the same state.

In this section we investigate these questions using negative binomial regression models. In each case, we fit models on appropriate 2003-2004 events that either represent the presence or absence of the proposed effect, and compare them in terms of both fit and predictive ability on 2005 data.

Consider first the question of leadoff single versus leadoff walk (or hit by pitch). The analysis is based on all at bats that lead off an inning. For the American League data, an initial examination implies that there is a difference between the outcomes of interest (walk, hit by pitch, single), but not in the anticipated direction: the average numbers of runs scored following a leadoff walk, hit by pitch, and single, respectively, are 0.84, 0.83, and 0.92, respectively. That is, an average of roughly 0.10 more runs are scored after a leadoff single than after a leadoff walk or hit by pitch. A similar pattern occurs when including situation effects predictors (since only leadoff events are considered, momentum effects are not relevant). The model that treats the three outcomes as different has lower *AIC* than the model that pools them as one event, but it is still leadoff singles that have the highest estimated runs scored given the other predictors. Even this effect is unimportant from a predictive point of view, however, as the simpler model that pools the outcomes is more accurate than the model that treats them separately 52.3% of the time when validated on the 2005 data. The situation in the National League is even more straightforward, as the model pooling the outcomes has smaller *AIC* than (and hence is preferred over) the model treating them separately. Thus, there is no meaningful evidence that leadoff walks, hit by pitches, or singles are any different in terms of how they lead to runs.

In order to investigate the rally-killing effect of the double play, we examine events where there are two outs and fewer than three baserunners (since a double play cannot result in three runners on base), and compare the runs scored from that point on when the previous state was zero outs (a double play) to those where the previous state was one or two outs. Once again a preliminary look at the data indicates an effect, and this time in the expected direction, as the average number of runs scored after a double play (0.16 in both the American League and the National League) is significantly lower than the average number scored when the previous play was not a double play (0.26 in the American League and 0.27 in the National League). When situational and momentum effects are taken into account the magnitude of the effect diminishes slightly, but in both leagues the model with a double play indicator is preferred by *AIC*. The validation performance on 2005 data is very similar for the two leagues, in that the model that includes the double play indicator provides more accurate predictions roughly 70% of the time, and roughly 80% of the time that no runs were scored after the play. Thus, there appears to be real evidence that a double play does lead to poorer offensive performance than would be implied by the current state and qualities of the pitcher and hitter, and this can provide meaningful gains in predicting scoring of future events.

8.6 CONCLUSIONS

In this paper we have examined the evidence for the existence of momentum effects in baseball. Model fitting and validation on new data demonstrate that in most cases, the evidence for momentum (in the sense that the results of previous plays affect the results of future plays) is weak, if not nonexistent. Although there is evidence that when one or two runners are on base, the result of the previous play affects the result of the next play and the number of runs scored from that play on in a real way, and that this effect mostly corresponds to batting failures (outs) following failures, as opposed to batting successes (such as hits) following successes, there is little evidence that this can be used to provide meaningful gains in predictive power on future data. Further, there is no evidence to support the often-heard belief that a leadoff walk is any worse for a pitcher than a leadoff single. On the other hand, the rally-stopping effect of a double play does appear to be real, and manifests itself in noticeably better predictions on new data.

REFERENCES

Albert, J. (2001). *Using play-by-play baseball data to develop a better measure of batting performance.* Bowling Green State University. Online at `bayes.bgsu.edu/papers/rating_paper2.pdf`.

Albert, J. (2003). *Teaching Statistics Using Baseball.* Washington, DC: Mathematical Association of America.

Albert, J. and J. Bennett (2003). *Curve Ball: Baseball, Statistics, and the Role of Chance in the Game* (revised ed.). New York: Springer-Verlag.

Albright, S.C. (1993). A statistical analysis of hitting streaks in baseball. *Journal of the American Statistical Association 88*, 1175–1183 (with discussion).

Burnham, K.P. and D.R. Anderson (2002). *Model Selection and Multimodel Inference: A Practical Information-Theoretic Approach* (2nd. ed.). New York: Springer-Verlag.

Howard, R.A. (1960). *Dynamic Programming and Markov Processes.* Cambridge, MA: MIT Press.

Koop, G. (2004). Modelling the evolution of distributions: An application to Major League Baseball. *Journal of the Royal Statistical Society, Series A 167*, 639–655.

Linhart, H. and W. Zucchini (1986). *Model Selection.* New York: John Wiley and Sons.

McQuarrie, A.D.R. and C.-L. Tsai (1998). *Regression and Time Series Model Selection.* Singapore: World Scientific.

Simonoff, J.S. (2003). *Analyzing Categorical Data.* New York: Springer-Verlag.

9

Inference about batter-pitcher matchups in baseball from small samples

Hal S. Stern

University of California, Irvine

Adam Sugano

University of California, Los Angeles

ABSTRACT

Baseball games involve a series of matchups between batters and pitchers. The outcomes of these matchups determine which team scores more runs and consequently which team wins the game. Data is available about the outcomes of the matchups, but for any given batter-pitcher combination the number of observed trials is quite small. The sample sizes are often below 20 and only rarely go above 75. There is a tendency for baseball fans and even baseball professionals to draw strong conclusions based on these small samples. For example, a player may be given a day off instead of playing against a pitcher for whom he has had little success in the past. Is this the right strategy? Or is this an overreaction to bad luck? This chapter considers a hierarchical beta-binomial approach to modeling the results of batter-pitcher matchups. The results suggest that there is some variability in the ability of a batter across different pitchers (or in the ability of a pitcher across different batters), but much less than one might think.

9.1 INTRODUCTION

One aspect of baseball that makes it a popular topic for statistical analysis is that the game is comprised of a series of individual batter versus pitcher matchups. These matchups are at the heart of the large data sets that are analyzed by baseball enthusiasts (e.g., Albright, 1993, James, 2006). The following anecdote indicates exactly how important information about these matchups can be in professional baseball. It was reported that Los Angeles Dodger player Kenny Lofton was rested by Dodger manager Grady Little during the team's August 29, 2006 game against the Cincinnati Reds because of Lofton's record against the scheduled Reds pitcher, Eric Milton. Lofton, whose season-long batting average (proportion of batting attempts resulting in base hits) was 0.308 at the time and whose career batting average was 0.299, had just one hit in 19 career attempts against Milton. This is a batting average of 0.056, considerably lower than Lofton's usual level of performance. Baseball players oc-

casionally take a night off during the season and to the manager this seemed like
the perfect night to give Lofton a rest. But is it? Lofton's substitute that night had
a batting average of 0.273 at the time, somewhat lower than Lofton's average. This
raises an important question about how a baseball manager should combine the limited information about a particular batter-pitcher matchup with the larger amount of
data available about the individual players. Some managers are reputed to believe
strongly in the importance of batter-pitcher matchups. In the book *Three Nights in
August*, author Buzz Bissinger tells us that St. Louis Cardinal manager Tony LaRussa
carries a card each game with the performances of his players against the pitchers of
the opposing team (Bissinger, 2005). There is certainly baseball logic that supports
considering such data. The way a certain pitcher releases the ball may make it easier for a particular hitter to see it, or a certain pitcher's tendencies (to throw mainly
fastballs for example) may match up well with a particular hitter's strengths. At the
same time however some basic statistics suggests that poor performance against one
pitcher may just reflect bad luck. After all there is about a $1/130 \approx 0.009$ chance that
a hitter of Lofton's ability would have one hit in 19 attempts just by chance. Lofton's
been around a long time and opposed many pitchers. Is he really having a tough time
with Milton? Or has he just been unlucky? In what follows we look more closely at
baseball pitcher-hitter matchups using a hierarchical beta-binomial model.

9.2   THE BATTER-PITCHER MATCHUP: A BINOMIAL VIEW

The simplest statistical approach to matchup data is to view the number of hits y in
a sequence of n trials between a particular batter and pitcher as a binomial random
variable with probability of success p. One can infer p from the observed data or
in the present case assess whether a particular data set is consistent with a particular
hypothesis about p. For the Lofton-Milton matchup the goal is to assess whether
the observed data ($y = 1$ in $n = 19$ trials) is consistent with $p = 0.300$ (Lofton's
approximate ability level based on career performance). The binomial distribution
with $n = 19$ and $p = 0.3$ yields $\Pr(y = 1) = 0.009$ and $\Pr(y \leq 1) = 0.01$. These
data suggest that a 1-for-19 run is unusual for a 0.300 hitter.

It is clear however that the binomial calculation omits an important consideration,
namely the multiplicity of possible matchups. It is unusual to find 1 success in 19
prespecified trials for an event with success probability 0.300, but it is not necessarily unusual to find 1 success in 19 trials if one considers many different sets of 19
trials before settling on the one to look at. In baseball terms we need to account for
the fact that just by chance Lofton will have a higher batting average against some
pitchers and a lower batting average against others. One way to learn about this is by
simulation. During the course of Lofton's career he faced many different pitchers. If
we take the 100 pitchers with whom Lofton has the most career matchups and simulate Lofton's performance assuming constant probability of success 0.300, then we
find it is not at all unusual to have one pitcher against whom Lofton's performance
is 1-for-19 or worse; in fact, this happened in 20% of the simulated careers. These
simulation results are consistent with other's findings as well. On-line baseball writer
Dan Fox gathered data on more than 30000 individual pitcher-batter matchups (Fox,

2005). His conclusion was that there is nothing in the data that one would not expect to see just by chance.

The simple binomial calculation and the simulation study suggest that there may be much less to batter-pitcher matchups than baseball fans and baseball managers believe. All of the observed data appear to be consistent with a simple binomial model. Such a conclusion however is almost certainly inappropriate. For one thing, the failure to reject a null model does not mean the null model is correct. The power to reject a null model or null hypothesis depends on the true size of the effect under study and the sample size. It seems likely that the variability of a batter's ability from pitcher-to-pitcher (the effect size) is not very big and the sample sizes are generally fairly small. These combine to make it difficult to reject the null model (see also Stern and Morris, 1993, for related work). In fact, one can be fairly certain that our approach is not powerful enough to detect failures of the simple binomial model; the reason for feeling this way is that there are some well-studied baseball phenomenona that tell us the simple binomial model is wrong. First, years worth of data across many, many players show that batters perform better against pitchers with the opposite dominant hand; in other words, left-handed batters have a higher average against right-handed pitchers than they do against left-handed pitchers (and vice versa for right-handed batters). This simple fact means a left-handed batter like Lofton should have a higher average against some pitchers (right-handed ones) than others. The second key fact is that some pitchers are better than other pitchers. Thus we expect at least some variation in success probability due to the quality of the opposing pitcher. Other factors may also be important including the site of the game (some stadiums are easier to hit in than others).

The two arguments presented so far, data analysis arguing in favor of the null binomial model and other information arguing against it, cause us to consider a hierarchical model that allows for a compromise. In essence, the hierarchical model that we introduce below allows the data to determine the degree to which the binomial model with constant probability holds. The approach is closely related to the ideas of shrinkage estimation described, for example, in Lehmann and Casella (2003), and demonstrated in the baseball context by Efron and Morris (1977).

9.3   A HIERARCHICAL MODEL FOR BATTER-PITCHER MATCHUP DATA

9.3.1   DATA FOR A SINGLE PLAYER

To make the discussion concrete Table 9.1 provides data for a number of batter-pitcher matchups involving Derek Jeter of the New York Yankees. During his 11 year career (through July 23, 2006), Jeter had faced 382 pitchers at least five times. The outcomes from a small number of those matchups are provided in the table along with his aggregate record, 2061 hits in 6530 attempts for a batting proportion, known as the batting average, of 0.316. Data were collected from the website www. sports.yahoo.com. The table shows extremely high and low batting averages based on a variety of sample sizes. Recent work on baseball statistics suggests that

TABLE 9.1

Derek Jeter's record against selected pitchers.

Pitcher	At-bats	Hits	Average
R. Mendoza	6	5	0.833
H. Nomo	20	12	0.600
A.J. Burnett	5	3	0.600
E. Milton	28	14	0.500
D. Cone	8	4	0.500
R. Lopez	45	21	0.467
K. Escobar	39	16	0.410
J. Wetteland	5	2	0.400
T. Wakefield	81	26	0.321
P. Martinez	83	21	0.253
K. Benson	8	2	0.250
T. Hudson	24	6	0.250
J. Smoltz	5	1	0.200
F. Garcia	25	5	0.200
B. Radke	41	8	0.195
J. Julio	13	0	0.000
D. Kolb	5	0	0.000
TOTAL	6530	2061	0.316

the batting average may not be the most reliable measure of hitting performance (see, for example, Chapter 7 by Albert in this book). We consider it here because it is still widely reported and more importantly because we could easily obtain the data we needed.

9.3.2 A PROBABILITY MODEL FOR BATTER-PITCHER MATCHUPS

The probability model specification begins with the simple binomial model for the outcomes in a single batter-pitcher matchup. Let n_i denote the number of batting attempts for Derek Jeter against pitcher i, and let y_i denote the number of hits (successful outcomes) where $i = 1, \ldots, I$, and I is the number of pitchers faced by Jeter. Assuming the at-bats against a pitcher can be treated as independent, identically distributed trials leads us to assume that $y_i \sim \text{Binom}(n_i, p_i)$ with p_i the probability of success for Jeter against the ith pitcher. One can question the iid assumption as the probability of success in a given matchup may depend on the site of the game, the score, the weather, etc. The model can be modified to accommodate such covariates (see, for example, Chapter 7 by Albert), but we do not consider them here primarily because we do not have that information available.

To address the possibility that Jeter's ability varies across opposing pitchers, we model the p_is as draws from a $\text{Beta}(\alpha, \beta)$ probability distribution. The beta distribution is commonly used as a probability model for parameters (like the batting average) that are known to lie in the interval from 0 to 1. The mean of the beta dis-

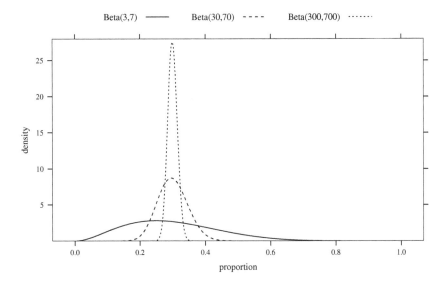

FIGURE 9.1
Three beta distributions with common mean $\alpha/(\alpha + \beta) = 0.3$ but varying degrees
of concentration about the mean.

tribution is $\mu = \alpha/(\alpha + \beta)$, and the variance is $v = \left(\frac{\alpha}{\alpha+\beta}\right)\left(\frac{\beta}{\alpha+\beta}\right)\left(\frac{1}{\alpha+\beta+1}\right)$. The
variance is proportional to the usual Bernoulli variance $(\mu(1-\mu))$ with the additional
term $\phi = 1/(\alpha + \beta + 1)$ reflecting the degree to which the beta distribution is concentrated around the mean. Large values of α, β (small values of ϕ) correspond to
distributions that are highly concentrated around the mean. Figure 9.1 illustrates the
beta distribution for three different choices of (α, β) having the same mean (0.3) but
varying degrees of concentration around that mean. The family of beta distributions
for the batter-pitcher matchup success probabilities provides a flexible family that
can accommodate data that appears to be consistent with constant success probability ($\phi = 0$) or highly variable success probability (larger ϕ). This joint probability
model (binomial distributions for the outcomes and a beta distribution for the binomial probabilities) is known as the beta-binomial hierarchical model. It has been
applied elsewhere with sports data (Morrison and Kalwani, 1993; Chapter 7). To
complete the probability model we require a prior probability distribution $p(\alpha, \beta)$
on the beta distribution parameters α and β. We select this prior distribution to reflect great uncertainty about the beta distribution parameters so that the inference
about the beta distribution will be determined to as large an extent as possible by the
data. Following the discussion in Gelman, Carlin, Stern, and Rubin (2003) we use a
flat prior distribution on the parameters μ and $(\alpha + \beta)^{-1/2}$ which is equivalent to
assuming $p(\alpha, \beta) \propto (\alpha+\beta)^{-5/2}$. Though this is not a proper probability distribution
it does lead to a proper posterior distribution.

TABLE 9.2
Derek Jeter's estimated ability against selected pitchers.

Pitcher	At-bats	Hits	Observed Average	Estimated Average	Posterior 95% interval
R. Mendoza	6	5	0.833	0.322	(0.282, 0.394)
H. Nomo	20	12	0.600	0.326	(0.289, 0.407)
A.J. Burnett	5	3	0.600	0.320	(0.275, 0.381)
E. Milton	28	14	0.500	0.324	(0.292, 0.397)
D. Cone	8	4	0.500	0.320	(0.278, 0.381)
R. Lopez	45	21	0.467	0.326	(0.292, 0.401)
K. Escobar	39	16	0.410	0.322	(0.281, 0.386)
J. Wetteland	5	2	0.400	0.318	(0.275, 0.375)
T. Wakefield	81	26	0.321	0.318	(0.279, 0.364)
P. Martinez	83	21	0.253	0.312	(0.254, 0.347)
K. Benson	8	2	0.250	0.317	(0.264, 0.368)
T. Hudson	24	6	0.250	0.315	(0.260, 0.362)
J. Smoltz	5	1	0.200	0.317	(0.263, 0.366)
F. Garcia	25	5	0.200	0.314	(0.253, 0.355)
B. Radke	41	8	0.195	0.311	(0.247, 0.347)
J. Julio	13	0	0.000	0.312	(0.243, 0.350)
D. Kolb	5	0	0.000	0.316	(0.258, 0.363)
TOTAL	6530	2061	0.316		

The Bayesian approach to inference requires that we calculate the posterior distribution of the parameters (α, β) and of the individual matchup parameters p_i. The posterior distribution is

$$p(\alpha, \beta, \{p_i, i = 1, \ldots, I\} | \{Y_i, i = 1, \ldots, I\}) \propto$$

$$(\alpha + \beta)^{-5/2} \prod_{i=1}^{I} p_i^{y_i} (1 - p_i)^{n-y_i} \prod_{i=1}^{I} \left(\frac{\Gamma(\alpha + \beta)}{\Gamma(\alpha)\Gamma(\beta)} p_i^{\alpha-1} (1 - p_i)^{\beta-1} \right)$$

The results provided throughout the remainder of the chapter are obtained via simulation from this posterior distribution. A simulation algorithm is used to generate samples from the posterior distribution. One thousand random samples from the posterior distribution are used to estimate the posterior median and a 95% central posterior interval for each parameter.

9.3.3 RESULTS - DEREK JETER

In applying this model to the data for Jeter's batter-pitcher matchups (all 382 matchups, not just the ones in Table 9.1), we find that the estimated posterior median of $\mu = \alpha/(\alpha + \beta)$ is 0.318 and a 95% posterior interval for μ is (0.310, 0.327). This is what we would expect since Jeter's career average is 0.316 based on more than 6000 at-bats. This speaks to Jeter's overall ability but then one can also obtain the posterior distribution for his estimated ability in particular matchups. Table 9.2 presents results for the matchups introduced in Table 9.1. The information from the initial table is

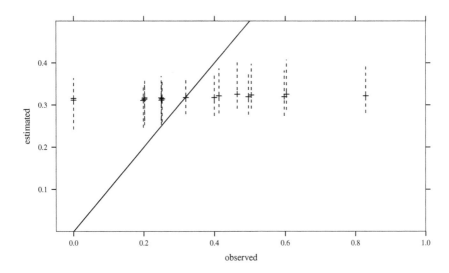

FIGURE 9.2
Plot showing the estimated batting ability of Derek Jeter against 18 different opposing pitchers from Table 9.1. Horizontal axis shows observed success probability. Plus
signs indicated median estimate of ability and dashed lines indicate 95% posterior
intervals for the ability. The 45-degree line is provided for reference.

repeated along with the posterior median and a 95% posterior interval for Jeter's
average in matchups with a particular pitcher. The most important thing to notice
is that Jeter's estimated ability is nearly constant! The highest estimated ability for
Jeter is 0.326 against Hideo Nomo against whom Jeter has 12 hits in 20 attempts. The
lowest estimated ability is 0.311 (just 0.005 below Jeter's career average of 0.316)
against Brad Radke against whom Jeter has 8 hits in 41 attempts.

A graphical display of the results is provided in Figure 9.2. The plus signs indicate the posterior median and the dashed lines indicate the posterior intervals for
Jeter's success probability against different pitchers. These are plotted against the observed average. The graph shows clearly that the observed heterogeneity in matchup
averages is discounted heavily in the estimated abilities. The estimates are said to
"shrink" away from the observed success probabilities in a given matchup towards
the overall success probability for Jeter. The amount of shrinkage depends on two
factors. First, it depends on how much heterogeneity is evident in the data and second
on the amount of data. The estimates for all of the high and low average matchups are
close to Jeter's lifetime average. This indicates that there is not much evidence for
heterogeneous performance. The effect of sample size can be seen in that the relative
order of the estimated abilities does not match the order of the observed averages.
The first matchup in the table which corresponds to a batting average of 0.833 is
discounted heavily because it is based on only 6 attempts. Another feature of the

graph that is striking is that the posterior intervals are extremely wide. Even with 83 attempts (the most for any matchup involving Jeter) there remains great uncertainty about the true ability of Jeter versus pitcher Pedro Martinez. (It should be noted that the posterior interval is narrower than it would be in a simple binomial analysis because the posterior interval combines information from the 83 at-bats with information from the beta distribution of Jeter's averages against other pitchers).

9.3.4   RESULTS - MULTIPLE PLAYERS

It is natural to wonder whether the pattern that we find above pertains to other hitters as well. We repeated the analysis for 230 hitters. These are batters with at least 350 plate appearances (indicating they were significant players) in 2006. The career data are analyzed for each player; the sample includes both young players and experienced players. A key parameter is $\phi = 1/(\alpha + \beta + 1)$ which indicates the estimated heterogeneity in a batter's ability across the population of pitchers. Figure 9.1 provides one way to calibrate the values of ϕ. The Beta(3,7) distribution has $\phi = 0.091$; the Beta(30,70) distribution has $\phi \approx 0.01$; the Beta(300,700) distribution has $\phi \approx 0.001$. The smaller the value the more concentrated the distribution and the less variable a batter's ability is across pitchers. For Derek Jeter the posterior median of ϕ is 0.002. (Note this is an estimate with uncertainty attached, but we are confident that ϕ is less than 0.01 for Jeter.) This value is quite small and reflects the fact that Jeter's ability varies a bit but not very much across the population of pitchers.

Over 230 major league batters we found the median value of ϕ to be about 0.005. Thus Jeter is more consistent than the typical players in the sense that his ability does not appear to vary much across pitchers. The values in this group of batters ranged from 0.0006 to 0.11. High values generally correspond to players with less information (young players) for whom performance to date suggests great variability. There is typically a great deal of uncertainty about the value of ϕ for such players.

Though ϕ is one useful way to characterize performance. It is not immediately obvious how ϕ translates into batting averages. Recall that Jeter ($\phi = 0.002$) had estimated abilities that ranged from about 0.311 to 0.326 across the 382 pitchers he had faced at least 5 times. To provide some context we note that Kenny Lofton (the player mentioned in the introduction) has estimated value of ϕ equal to 0.008. Though small, this value is considerably bigger than Jeter's value. The bigger value of ϕ is reflected in the estimates for Lofton's ability across the pool of pitchers. Lofton's estimated ability varies from 0.265 (against Eric Milton – the pitcher in the introduction) to 0.340. That is much more variability than we see in the estimates of Jeter's performance level.

9.4   BATTER-PITCHER DATA FROM THE PITCHER'S PERSPECTIVE

Most often one hears about batter-pitcher matchup data from the batter's perspective. As in the introduction to this chapter a player may sit out a game when the opposing pitcher is a bad matchup or alternatively may be removed from a game at a key juncture if there is a bad matchup. On the other hand most pitchers (including all starting pitchers) are put in the game to face more than one batter, and therefore managers are unlikely to make a change because of one matchup. There are exceptions to this, especially for relief pitchers at key points late in a close game, but we focus here predominantly on starting pitchers for whom the matchup sample sizes tend to be larger. There is nothing in the hierarchical model that is particular to the batters. The exact same model can be used to analyze batter-pitcher matchups from the pitchers' perspective. Is there any evidence that pitchers do better against some hitters than others?

9.4.1 RESULTS - A SINGLE PITCHER

To demonstrate the analysis for a pitcher consider the record of Derek Jeter's team-mate Mike Mussina. Mussina has been pitching for 16 years and through July 23, 2006 had faced 576 batters 5 or more times. The first three columns of Table 9.3 provides Mussina's results against a selection of hitters along with his career totals through that date. As with Jeter we focus only on batting average here though the analysis could be repeated for other measures of pitching effectiveness. Mussina's career record is excellent. He allows batters a success proportion of 0.250 which is below the average for the entire league. There is considerable variability in the outcomes for different hitters ranging from Mueller who has 0 hits in 23 attempts to Hidalgo who has 7 hits in 11 attempts. What does the hierarchical beta-binomial model suggest about Mussina's performance across the pool of batters?

The right hand side of Table 9.3 summarizes the posterior distribution of the probabilities of getting a hit for each player in the table when facing Mike Mussina. The results show similarities and differences relative to the analysis of the Jeter data. Once again the estimated abilities are concentrated much more closely around Mussina's career average 0.250 than the observed averages. The amount by which an individual estimate changes relative to the observed average depends on the sample size. Thus Mueller's lack of success in 23 attempts is reflected in a much lower estimated success probability (0.214) than the estimate for Kent (who has had no hits but only six attempts thus far). There is also considerable uncertainty about the true ability of each batter based on the relatively small sample sizes. The posterior intervals are quite wide.

The key difference relative to the data from Jeter is that there is more variation in the estimated abilities. They range from 0.214 (Mueller) to 0.312 (Catalanotto) which is a much bigger range than was evident for Jeter or other hitters. Part of the explanation lies in the selected individuals. Jeter is among the more consistent hitters – there is little variability in his success probability across pitchers (recall that ϕ which measures this consistency was 0.002 for Jeter). Mussina's performance is above the median in terms of heterogeneity of performance. The estimate of ϕ for Mussina is 0.008 which means the estimated abilities will be more spread out. Interestingly the estimated abilities for Mussina are more spread out than those for Kenny Lofton, a batter with a similar estimate for ϕ.

TABLE 9.3
Mike Mussina's record and estimated ability against selected hitters.

Pitcher	At-bats	Hits	Observed Average	Estimated Average	Posterior 95% interval
R. Hidalgo	11	7	0.636	0.278	(0.211, 0.368)
A. Cintron	5	3	0 .600	0.263	(0.191, 0.348)
B. Roberts	26	14	0.538	0.296	(0.231, 0.389)
R. Ibanez	21	10	0.476	0.279	(0.216, 0.367)
F. Catalanotto	56	26	0.464	0.312	(0.248, 0.397)
R. White	11	5	0.455	0.265	(0.197, 0.352)
M. Huff	5	2	0.400	0.256	(0.185, 0.340)
F. Thomas	78	30	0.385	0.299	(0.239, 0.374)
P. Burrell	10	3	0.300	0.253	(0.184, 0.335)
J. Canseco	61	18	0.295	0.264	(0.205, 0.332)
B.J. Surhoff	40	10	0.250	0.250	(0.189, 0.319)
A. Soriano	8	2	0.250	0.250	(0.180, 0.331)
H. Baines	35	7	0.200	0.240	(0.175, 0.308)
T. Hafner	10	2	0.200	0.247	(0.178, 0.324)
C. Fielder	42	7	0.167	0.231	(0.166, 0.296)
S. Posednick	6	1	0.167	0.247	(0.175, 0.325)
B. Mueller	23	0	0.000	0.214	(0.138, 0.279)
J. Kent	6	0	0.000	0.240	(0.166, 0.316)
TOTAL	11954	2992	0.250		

9.4.2 RESULTS - MULTIPLE PLAYERS

The analysis of Mussina was replicated on career data for 158 other pitchers. These are all pitchers who pitched more than 85 innings during 2006. This includes both young pitchers for whom 2006 is their first year and more experienced pitchers (like Mussina). The distribution of ϕ (our measure of consistency or concentration) is quite similar to what was found in the population of hitters. The values range from a low of 0.0007 (which suggests almost no variability across batters) to 0.03 (which suggests considerable variability). Successful pitchers are found at both ends of the distribution. As remarked above Mussina is slightly above the median (which is 0.005).

One feature that is currently unexplained in our data is that it appears the distribution of individual batter-pitcher matchup ability estimates tends to be more spread out for pitchers than for hitters, even given the same value of ϕ. This is a bit paradoxical but may reflect the distribution of sample sizes.

9.5 TOWARDS A MORE REALISTIC MODEL

The analyses done here provide one natural check of the methodology. Derek Jeter had batted against Mike Mussina 33 times before they became teammates. Jeter has 12 hits for an everage of 0.363. This is above Jeter's typical average (0.316) and substantially higher than what Mussina usually allows (0.250). Based solely on the analysis of Jeter's record one finds that an estimate of Jeter's ability against Mussina is 0.319 with a 95% posterior interval of (0.279, 0.372). Based on the analysis of Mussina's record one finds that an estimate of Jeter's ability against Mussina is 0.271 with a 95% posterior interval of (0.210, 0.353). The point estimates differ considerably, but the posterior intervals are sufficiently wide that differences of this magnitude are not terribly unusual.

Though focusing on an individual hitter or pitcher seemed like a natural first step in addressing the question of batter-pitcher matchups it is disappointing to end up with two estimates for the same matchup that differ by so much. It is natural to wonder whether a unified modeling approach can provide better inferences in this setting. Work in this direction is ongoing, however, some preliminary ideas are presented here. In a unified approach one would likely model p_{ij} the probability of success for batter i against pitcher j. A logistic model might include terms for batter ability and pitcher ability together, e.g.,

$$\text{logit}(p_{ij}) = \mu + a_i + b_j$$

with the batter effects a_i and pitcher effects b_j given appropriate population distributions. Of course the question about whether matchups are important is really the question about whether an interaction is needed in the above model. There are various ways to incorporate an interaction. One positive feature of a more sophisticated modeling approach is that it naturally extends to allow consideration of other factors known to effect batting success like the ballpark and the dominant-handedness of the players. Such factors have been ignored in our study.

One interesting note is that baseball enthusiasts have themselves discovered a method for predicting the outcome of an at-bat between a batter with batting average *BA* and a pitcher with batting average allowed *PA*. Assuming the league average is *LA* the "log5" method developed by writer Bill James in his 1983 Baseball Abstract suggests the predicted average be taken as $\hat{p} = \frac{BA \times PA}{LA} / \left(\frac{BA \times PA}{LA} + \frac{(1-BA)(1-PA)}{(1-LA)} \right)$. Though unusual in this form some manipulation yields $\text{logit}(\hat{p}) = -\text{logit}(LA) + \text{logit}(BA) + \text{logit}(PA)$ which is remarkably similar to the logit model proposed above (with $a_i = \text{logit}(BA), b_i = \text{logit}(PA),$ and $\mu = -\text{logit}(LA)$).

9.6   DISCUSSION

The main result here is not particularly surprising to statisticians. The evidence is that baseball fans and baseball professionals may be over-reacting to the chance outcomes in small samples. A comprehensive analysis suggests that there is in fact much less variation in batting performance across different pitchers than would be suggested by looking at the results of small samples. For Derek Jeter our results suggest that his "true" ability against a pitcher rarely deviates from his overall ability (lifetime average) by more than about 0.010. When a batter has had success (or lack of success) against a particular pitcher in 5, 10, or even 20 or 50 at-bats we would be wise not to put too much stock in the data. That said there is also evidence that some batters are more heterogeneous in their batting ability than others. Jeter is among the most consistent. For Kenny Lofton, the batter whose experience started the chapter, there is evidence of more variation from pitcher to pitcher. In particular Lofton's estimated success rate against the pitcher that August day is estimated as being almost 0.060 lower than his lifetime average 0.299. Though still a modest difference this might be large enough to justify trying a different hitter.

Before closing though it is important to recognize that there are significant issues, both statistical and nonstatistical, that limit the inferences we should draw from the results presented here. On the statistical side we have worked primarily with data from a single batter and a single pitcher. Though summary results are provided for a range of other hitters and pitchers these have not been explored in as much detail. Perhaps more importantly, we would be wise to recall the words of the statistician George Box who said, "All models are wrong, some models are useful." The simple hierarchical model that we have used here is clearly not quite right. It ignores some of the factors, site of game and dominant hand information are two, which are known to impact the probability of a hit. When asking whether there are "matchup effects" we likely want to know whether there is anything beyond the known determinants of batting success that is affecting a particular matchup. The current analysis does not really address this question because it ignores some factors that we know are relevant. On the other hand taking account of such factors is likely to make matchup effects appear even less important than shown here.

Another caveat before criticizing the work of baseball decision makers is that non-statistical information is likely to impact a decision as well. In the situation that started the article, we are not privy to all of the information that Grady Little had nearby as he built his lineup for the August 29 game against Cincinnati. Perhaps Kenny Lofton was slightly injured or perhaps there was a player returning from the injured list whom he wanted to evaluate. The challenge of the manager is to balance quantitative information about the ability and expected performance of a player with factors like these that are less statistical.

Despite the statistical and nonstatistical caveats, the results here argue strongly against drawing conclusions based on the limited information available about the matchup between a particular batter and a particular pitcher. A manager should avoid reacting to small sample variation which may lead him to rest a superior player at an inopportune time.

REFERENCES

Albright, S.C. (1993). A statistical analysis of hitting streaks in baseball. *Journal of the American Statistical Association 88*, 1175–1183 (with discussion).

Bissinger, B. (2005). *Three Nights in August: Strategy, Heartbreak, and Joy Inside the Mind of a Manager.* Mariner Books.

Efron, B. and C. Morris (1977). Stein's paradox in statistics. *Scientific American 236*(5), 119–127.

Fox, D. (2005, November 10). Tony LaRussa and the search for significance. http://www.hardballtimes.com/main/article/tony-larussa-and-the-search-for-significance.

Gelman, A., J.B. Carlin, H.S. Stern, and D.B. Rubin (2003). *Bayesian Data Analysis* (2nd ed.). Boca Raton: CRC Press/Chapman & Hall.

James, B. (2006). *The Bill James Handbook 2007.* Skokie, IL.: ACTA Sports.

Lehmann, E.L. and G. Casella (2003). *Theory of Point Estimation* (2nd ed.). New York: Springer.

Morrison, D.G. and M.U. Kalwani (1993). The best NFL field goal kickers: Are they lucky or good? *Chance 6*(3), 30–37.

Stern, H.S. and C.N. Morris (1993). Looking for small effects: power and finite sample bias considerations. (Comment on C. Albright's "A statistical analysis of hitting streaks in baseball"). *Journal of the American Statistical Association 88*, 1189–1194.

10

Outcome uncertainty measures: how closely do they predict a close game?

Babatunde Buraimo

University of Central Lancashire

David Forrest

University of Salford

Robert Simmons

Lancaster University

ABSTRACT

A maintained hypothesis in modelling of a sporting contest is that attendance or audience varies positively with the expected closeness of the contest; this is the uncertainty of outcome hypothesis. The literature on attendance in team sports has used a number of proxies for ex ante outcome uncertainty, variously derived from league tables or betting odds. We test whether these proxies in fact have predictive power for the closeness of a match, something not previously attempted. We find that all the measures we consider have statistically significant but nevertheless very limited power. Betting odds based measures are superior but using odds to forecast closeness in results still fails to account for more than a small proportion of the variation in match outcomes from a large sample of Spanish soccer fixtures. We conclude that employment of such measures in modelling audience size is inadequate for the testing of impacts of outcome uncertainty, and it is unsurprising that the empirical literature yields no clear cut evidence in support of the hypothesis.

10.1 INTRODUCTION

The role of outcome uncertainty in increasing spectator appeal is a central focus of the academic literature on the conduct and regulation of professional sports leagues. If the result of a match is hard to predict, in that the two teams have similar ex ante probabilities of winning, then a close contest is likely and, because they value suspense, extra spectators are liable to be attracted to the stadium, raising club and league revenue.

This simple hypothesis has been at the heart of the defense case in many court proceedings, in America and Europe, where competition authorities have challenged restrictive business practices such as collusion in the selling of television rights or limitations on the freedom of movement of players. Very often the defense has been successful, courts accepting that special measures are necessary to equalize financial, and therefore, playing resources across clubs. This in turn, leagues have argued, will promote outcome uncertainty, ensure closely fought games and therefore attract more spectators.

But do evenly balanced matches in fact attract greater audiences? In the large literature reporting attempts to build regression models of attendances at matches in professional sports leagues, a measure of outcome uncertainty is typically included as one of the covariates. Borland and Macdonald (2003) tabulate the results of 18 such match-level studies (in various sports and on different continents) and reveal very mixed results, with only three cases where the hypothesis, that the size of crowd is affected by outcome uncertainty, is supported. Szymanski (2003) likewise describes the evidence as "far from unambiguous." Out of 22 cases cited by Szymanski, only ten offer clear support for the outcome uncertainty hypothesis.

For such a popular hypothesis, there is, then, curiously little empirical support. It might be concluded that, within the range of outcome uncertainty actually observed in sports leagues, variations in it from game to game are not in fact important to spectators considering whether or not to attend. The defense case in competition proceedings against sports leagues would then look very thin.

There is, however, an alternative possibility (unexplored in previous studies) to account for the lack of explanatory power of outcome uncertainty measures in crowd size models. This is that the proxies for outcome uncertainty used by the various authors typically fail to capture accurately the likelihood that a particular game will be close. Our starting point, in the spirit of the rational expectations paradigm (whereby the average forecast among a large number of people tends to be as correct as it could be given available information), is that sports fans in the aggregate are likely to have good intuition as to how closely contested a particular game is likely to be (if indeed this is capable of being forecast at all). A valid measure of outcome uncertainty (that will capture the behavior of potential spectators) will then be one that is effective as a predictor of a close game. The attendance literature features several different measures for ex ante outcome uncertainty but none of the studies report testing whether the proxy used is in fact highly correlated with match outcomes.

In this chapter, we will review the several measures of outcome uncertainty found in statistical analyses of attendance. For each, we will test whether it is actually an effective forecasting tool, where the object is to forecast closeness of contest. Further, we will assess whether any one measure outperforms the others to an extent such that findings based on it should be accorded greater weight when reviewing the existing attendance literature. Our work may also provide guidance to future researchers as to the most suitable outcome uncertainty measure for adoption in statistical modelling of crowd size.

10.2   MEASURES OF OUTCOME UNCERTAINTY

There is no consensus (or even debate) on what statistic might best capture the extent to which a match is likely to be closely contested, but the many indices adopted by various authors fall essentially into just two groups. The first comprises those which employ information found in league tables (as compiled immediately prior to a match) and the second those that betting odds as the basis for evaluation.

In their pioneering model of attendance in football, Hart, Hutton, and Sharot (1975) included as their uncertainty of outcome variable the absolute difference in the league positions of the home and away teams, as shown in the standings immediately prior to the game. The hypothesis was that the greater the discrepancy between the teams, the less likely the game was to be close and therefore the lower the size of the crowd would be (given the values for other covariates such as the distance away fans would have to travel). There was weak support for the hypothesis though the data analyzed related to a small sample of attendance data at just four English football clubs. However, the measure proved insignificant when applied to a later (and larger) data set from the English Premier League by Baimbridge, Cameron, and Dawson (1996). It could be, of course, that English Premier League crowds are in fact insensitive to outcome uncertainty, but an alternative explanation is that the measure itself fails to capture whether the teams are likely to prove well matched on the day. To illuminate the issue, our first measure of outcome uncertainty to be tested is therefore:

$$X_1 = \mid \text{home league position} - \text{away league position} \mid. \qquad (10.1)$$

A weakness of using absolute difference in league position (or in league points per game achieved in the season to date) to proxy outcome uncertainty is that it fails to account for the phenomenon of home advantage, which is significant in most, if not all, team sports. For example, in most professional football leagues, approximately twice as many matches are won by home as by away teams. Therefore, if the value for X_1 is, for example, seven, most commentators would expect a one-sided game if the home team occupied the superior league position (since it enjoys both greater strength and home advantage) but a close game if the home team trailed by seven places (since then its lesser strength could be compensated by the advantage of playing in its own stadium). Paton and Cooke (2005), in a study of attendances at English cricket matches, and Forrest, Simmons, and Buraimo (2005), modelling the size of television audiences in football, proposed that any measure of outcome uncertainty based on information (position or points) in league tables should incorporate an adjustment for home advantage. Here, we test the efficacy of the Forrest et al. (2005) measure (which was significant in their context) in forecasting closeness of contest:

$$X_2 = \mid \text{home advantage} + \text{home points-per-game} - \text{away points-per-game} \mid \quad (10.2)$$

where home and away points are calculated for the season up to the time of the match and where home advantage = mean points-per-game achieved by all home teams in the previous season *minus* points-per-game achieved by all away teams in the previous season. This measure X_2 may be considered a refinement of X_1 because it both corrects for home advantage and exploits a cardinal (points) rather than an ordinal (position) summary statistic for team strength.

The number of potential outcome uncertainty measures based on information in league tables is large since position, wins, or points may be employed and any of these may be transformed, for example by squaring the value. However, we take X_1 and X_2 as representative of this general class of measures, and below we test how well they forecast close games.

Another set of authors, from Peel and Thomas (1988) on, has turned to the betting market for indicators of outcome uncertainty to employ in modelling crowd size. Other examples in a football context, used with mixed results, include Peel and Thomas (1992), Kuypers (1996), and Czarnitzki and Stadtmann (2002). Rascher (1999) and Welki and Zlatoper (1999) similarly proxied uncertainty of outcome by reference to betting market information (odds or point spreads) in attendance studies for baseball and American football.

Of course, growing interest in sports betting has generated many styles of bet, so that odds in the case of football are available on a variety of dimensions of a game, for example exact score or goal supremacy. But the authors investigating outcome uncertainty in soccer have used odds quoted in respect of match outcome, in the sense of home win, draw, or away win. Let the probabilities of match outcomes implied by these odds* be ϕ_h, ϕ_d, and ϕ_a respectively. Peel and Thomas used ϕ_h as the outcome uncertainty measure, but interpretation of model results is difficult because uncertainty is greater for middle than for either high or low values of the variable. The implied probability of a draw, ϕ_d, refers to the closest result of all; but because of the way that the industry sets its odds, this shows very little variation across matches, for example in the data set we analyze below, the standard deviation of draw odds (quoted as implied probabilities) is 0.027 compared with 0.123 for home odds and 0.106 for away odds.

The most natural odds-based measure to capture uncertainty appears to us to be the difference between the odds quoted on either team winning the match. Accordingly our third candidate variable tested for efficiency in forecasting closeness of contest is:

$$X_3 = |\phi_h - \phi_a|. \qquad (10.3)$$

The arguments in favor of using betting odds as the basis for uncertainty measures, in preference to league positions or points, are strong. Bookmakers have an obvious financial incentive to set accurate odds to the extent that, especially given falling margins in the era of strong competition for internet wagering, any errors may result in providing professional bettors with opportunities for positive expected returns. The

---

*For example, suppose odds quoted according to UK convention were 3:1 (3 units profit for one unit stake). They would be quoted as 4.00 in most of Europe (collect 4 units including return of stake for a one unit stake). The corresponding "probability-odds" would be 1/4 or 0.25. But the sum of these probability-odds will be greater than one to ensure bookmaker profit. Suppose they summed to 1.12. "Implied probability," ϕ, would then be noted as 0.25/1.12 or 0.223, i.e., in the calculation of "implied probability," probability-odds are scaled so that they add up to one.

ability of odds-setters to process information effectively has been documented in academic studies which illustrate that their forecasting performance can better those of expert commentators (Boulier and Stekler, 2003) and statisticians (Forrest, Goddard, and Simmons, 2005). Since these expert odds-setters can employ information in the league tables but use additional information, for example on player suspensions and injuries, it would appear likely that odds would provide a more reliable signal on the prospective closeness of a match.

On the other hand, odds-setters are not employed to generate odds that serve as forecasts but rather to provide odds consistent with maximization of profit. They may therefore consciously introduce biases into posted odds to exploit misperceptions or preferences among bettors. For example, Forrest and Simmons (2002) found evidence for English club football that more generous odds were typically offered in respect of wagers that more heavily supported teams would win. Contamination of odds by such "sentiment" may reduce their value in forecasting, for example, closeness of contest. Forrest and Simmons proposed measuring outcome uncertainty with reference not to actual odds but to odds adjusted for biases identified in preliminary modelling of the betting market. When they replaced actual by adjusted odds to capture the role of outcome uncertainty in a model of football attendance, they found that their measure then became statistically significant, underlining the potential sensitivity of findings to the choice of proxy. Our fourth measure of outcome uncertainty to be tested is therefore:

$$X_4 = |\phi_h^{adj} - \phi_a^{adj}| \tag{10.4}$$

where the ϕ this time refer to home and away probabilities adjusted for any biases identified in the betting market.

10.3   DATA

The context for our statistical analysis is the premier division of Spanish football, selected as a case study for the ready availability of sporting and betting data, for the strong interest it attracts throughout Europe and for the fact that there is a wide variation in resources available across clubs (with three dominant teams, Barcelona, Real Madrid, and Valencia). Outcome uncertainty is therefore at least as relevant an issue here as it is in any of the other major leagues which were open to study.

We collected data for both match results and betting odds from football-data.co.uk.[†] The source includes odds quoted by several internationally reputable bookmakers, which are naturally very closely correlated with each other. We chose to use those from William Hill because it had the largest number of matches where odds were quoted (i.e., fewest missing values). League tables required for constructing measures X_1 and X_2 were generated from a program we wrote for the purpose.

Our data set for analysis relates to seasons 2000/01 to 2004/5, with the following season, 2005/06, retained as a hold-out sample. Because our measures to be tested

---

[†] Access to this service requires payment of a small subscription.

included reference to league table positions, we excluded the first three rounds of matches from each season since there is no league table until after the first round and this first set of standings each year will unduly reflect, for example, which teams happen to have been scheduled to play at home in week one. There were also missing observations where odds were not recorded, allowing us finally to work with a sample of 1606 matches.

10.4   PRELIMINARY ANALYSIS OF THE BETTING MARKET

Whereas our measure X_3 is underpinned by an assumption of efficiency in the betting market, measure X_4 is based on the contrary notion that betting odds will be systematically biased and that these biases need to be removed before implied outcome probabilities can be taken as forecasts. Biases may take a number of forms. For example, it is well known that horse betting markets display "longshot bias" such that short (long) odds outcomes are even more (less) likely to occur than the odds imply. Another type of bias may be termed "sentiment bias": odds may reflect the preferences of bettors, for example which team they want to win, as well as objective factors relevant to match outcome. Forrest and Simmons (2002) found evidence of such bias in English football where bookmakers appeared to offer more generous odds in respect of wagers on more heavily supported teams.

To construct X_4 for matches in our Spanish data, we first estimated an ordered probit model where match result (0 = home win, 1 = draw, 2 = away win) was regressed on ϕ_h (implied home win probability) and on a variable *diffattend*, intended to represent the difference in the number of supporters of each team.[‡] The equation estimated was:

$$\text{Outcome} = -1.66\phi_h - 0.0000056\,diffattend \qquad (10.5)$$
$$(-4.15) \quad\ (2.69)$$

where z-statistics appear in parentheses, and $n = 1606$.

Given that home win is coded zero, the negative sign on the odds variable, ϕ_h, indicates that a home win becomes more likely as odds on a home win shorten. This is as one would expect. However, if the market were efficient, the odds would capture all information relevant to match outcome and additional variables would be redundant in the model. But, in fact, the variable *diffattend* is also strongly significant and its negative sign implies that "bigger" teams are more likely to win than the odds alone would suggest, i.e., odds are more generous in respect to bets on "big" teams than bets on "small" teams. This is confirmation for Spanish soccer of the result reported for the English context by Forrest and Simmons. This evidence of betting market inefficiency suggests it is appropriate to derive adjusted probabilities

---

[‡]*diffattend* equals the mean home attendance of the home team in the previous season minus the mean home attendance of the away team in the previous season. In a high profile league such as the one we study, passive fans may outnumber those who attend the stadium by a large margin, but we take attendance as an adequate proxy for a club's aggregate level of support. In our data set, *diffattend* sometimes takes a value in excess of 60000, reflecting strong disparity in size of club across the division.

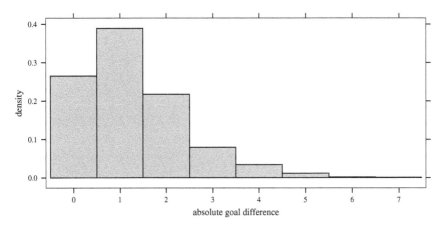

FIGURE 10.1
Absolute goal difference in results of 1606 Spanish football matches.

for home and away wins, and we did so using the estimated ordered probit relationship. The flexible shape of the relationship estimated in ordered probit permits the procedure to account for any longshot bias in the odds and the presence of *diffattend* adjusts for sentiment bias. Outcome uncertainty measure X_4 is calculated from the adjusted values of ϕ_h and ϕ_a generated as fitted values by the probit equation.

With all the outcome uncertainty variables, X_1 to X_4 constructed, we now test their relative performance in accounting for the closeness of a match.

10.5 MODEL

We seek to account for how close matches are, measured by the absolute goal difference between the teams at the end of the game. This gives us count data with a distribution shown by the histogram presented as Figure 10.1. The dominant outcomes are those in which the match is either tied (goal difference = 0) or where one goal separates the teams. But absolute goal difference ranges as high as 7 and its mean is 1.275.

These are count data with a nonnormal distribution, and therefore the Poisson model is a candidate in the choice of model to be estimated. If a linear regression model were applied to our data set by ordinary least squares, the results might be inefficient, inconsistent, and biased. The Poisson model is then more reliable where the data consist of counts. Count data models have been employed by a number of authors as a basis for generating forecasts of the number of goals scored by a team in a football match, see, for example, Dixon and Coles (1997).

The properties of the Poisson model include equidispersion, i.e., mean = variance. In our data set, the mean and variance of absolute goal difference are in fact very close, 1.29 and 1.32 respectively. The Poisson distribution therefore appears to be a reasonable approximation to the observed distribution, and we proceed to es-

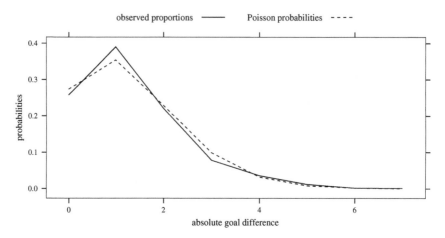

FIGURE 10.2
Observed and predicted proportions of absolute goal difference using X_1.

timate with a Poisson regression model. If the variance exceeds the mean, we have overdispersion; the Poisson model underestimates the amount of dispersion in the dependent variable. In such a situation, the negative binomial regression model can be applied which adds a parameter, α, that reflects unobserved heterogeneity among the observations (Long and Freese, 2003). In our case, we test for overdispersion with the null hypothesis that $\alpha = 0$. *Stata* 9.0 offers a likelihood ratio test of this hypothesis, and the p-value from a negative binomial regression model was always greater than 0.30 across the measures of outcome uncertainty considered. The null hypothesis of equidispersion is not rejected and the negative binomial model reduces to the Poisson regression model.[§] Hence, we report Poisson regression results below.

We estimated the model for each of the four alternative measures of outcome uncertainty and, each time, the distribution of predicted values of absolute goal difference came close to replicating the observed distribution, again validating our choice of Poisson (see, for example, Figure 10.2 which shows the predicted and observed distributions in the case of uncertainty measure, X_1).

We estimated the relationship between absolute goal difference and, in turn, each of our outcome uncertainty measures. The results are displayed in Table 10.1.

All measures succeed in capturing information relevant to how close a match will be to the extent that X_i is always strongly significant. Assessing the relative performance of each measure by reference to the values of log likelihood and pseudo R^2, the variables X_3 and X_4, derived from betting odds, yield greater goodness of fit than those, X_1 and X_2, based on league tables. This vindicates the greater popularity of

---

[§]By contrast, if we were modelling actual rather than absolute value of home goals minus away goals, there would be more zeros than would be predicted by Poisson and, with such "overdispersed" data, it would then be appropriate to employ a negative binomial regression model.

TABLE 10.1

Poisson regression results (dependent variable: absolute goal difference, $n = 1606$; z-statistics in parentheses).

	Constant	Coefficient on X	Log likelihood	Pseudo R^2
X_1	0.182 (4.46)	0.011 (2.26)	−2328.48	0.001
X_2	0.131 (3.18)	0.164 (3.69)	−2324.29	0.003
X_3	0.150 (4.80)	0.504 (5.15)	−2317.66	0.006
X_4	0.135 (4.10)	0.507 (5.34)	−2316.67	0.006

betting market proxies in the more recent literature relating crowd size to outcome uncertainty. Within each category, the more refined measure improves fit, so that allowing for the phenomena of home advantage and biases in betting odds is worthwhile. However, notwithstanding the strongly significant bias in odds identified in preliminary modelling, the improvement from making an appropriate correction to the odds variable is very marginal.

While each outcome uncertainty measure is shown to account to some extent for the score difference in a match, its role is nevertheless very limited, even for the "best" measure, X_4, as is reflected in the low values for pseudo R^2: "noise" accounts for the bulk of the variation in absolute goal difference across matches. This does not necessarily mean that it would be impossible to model the closeness of a game with a sufficiently data-rich statistical model; it is an open question whether the noise comes from true randomness of events or from observable variables excluded from our models. But the simple proxies for outcome uncertainty actually employed in the attendance modelling literature and examined here appear insufficiently correlated with match outcome for them convincingly to carry the weight of representing potential spectators' expectations over whether a match will be close. This gives a clue to the failure of outcome uncertainty to play a statistically significant role in most published models of attendance at team sports events.

10.6 OUT-OF-SAMPLE TESTING

So far, the ability of outcome uncertainty measures to predict match results has been assessed for five seasons. We now proceed to evaluate the forecasting ability of these measures for 370 games in the 2005/06 season. Table 10.2 shows correlations of predicted and actual values of absolute goal difference, with predicted values derived from each of our outcome uncertainty measures X_1 to X_4.

The reported correlation coefficients are rather low, in line with the small values of pseudo-R^2 in the main equations. The ranking of correlations matches those of log-likelihood and pseudo-R^2 in the main model. Essentially, the measures using betting odds outperform those using league positions or points. Hence, the results from the simple forecasting exercise corroborate the findings from the main model.

TABLE 10.2
Correlations of predicted and actual absolute goal
difference 2005/06

Outcome uncertainty measure	Correlation coefficient
X_1	0.054
X_2	0.042
X_3	0.116
X_4	0.121

10.7 CONCLUDING REMARKS

The chapter began with reference to a popular hypothesis, that the live (or television) audience for a match is a function of the degree of uncertainty concerning its outcome. This hypothesis is in one sense untestable because outcome uncertainty is an ex ante concept that is not directly observed. Researchers have to adopt proxies for it that they suppose represent adequately the ex ante probability distributions over match outcome in the minds of potential attendees or viewers. But these proxies cannot be tested for efficiency even if the audience is portrayed as able intuitively to settle on "correct" probability distributions because, when the match takes place, what is observed is not a probability distribution of outcomes but only a single outcome.

We overcome this problem by reformulating the hypothesis to imply that the audience is interested not in outcome uncertainty per se but in seeing a closely fought game. High outcome uncertainty is taken to mean a high chance of a close contest. Now it is possible to test the efficacy of the outcome uncertainty proxies to the extent that the proxies can be compared with a measure of actual closeness of contest. We consider one readily observable measure of closeness of contest, namely absolute difference in goals scored by each of the two teams.

From the literature on crowd size have emerged a number of proxies for ex ante outcome uncertainty, variously derived from league tables or betting odds. We test whether these proxies in fact have predictive power for the closeness of a match, something not previously attempted. We find that all the measures we consider have statistically significant but nevertheless very limited power. Betting odds based measures are superior but using odds to forecast closeness in result still fails to account for more than a small proportion of the variation in the data from Spain that we have analyzed. We conclude that employment of such measures in modelling audience size is therefore inadequate for the testing of the outcome uncertainty hypothesis, and it is unsurprising that the empirical literature yields no clear-cut evidence. In light of our findings, this failure in the literature is consistent with three possible explanations. First, spectators might be indifferent to outcome uncertainty in the range of uncertainty actually present in professional sports leagues.[¶] Second, spectators might be able to distinguish between matches likely to be close and those likely to be one sided but their views will not be captured by simple proxies for outcome uncertainty because these are only weakly correlated with actual results. Third, spectators may recognize that noise dominates the determination of individual match results and that, while they hope for a close game, it is fruitless to choose which games to attend by trying to assess outcome uncertainty. Any of these explanations could account for why a large literature fails to support a relationship between attendance and outcome uncertainty.

REFERENCES

Baimbridge, M., S. Cameron, and P. Dawson (1996). Satellite television and the demand for football: A whole new ball game? *Scottish Journal of Political Economy 43*, 317–333.

Borland, J. and R. Macdonald (2003). Demand for sport. *Oxford Review of Economic Policy 19*, 478–502.

Boulier, B. and H. Stekler (2003). Predicting the outcomes of National Football League games. *International Journal of Forecasting 19*, 257–270.

Czarnitzki, D. and G. Stadtmann (2002). Uncertainty of outcome versus reputation: Empirical evidence for the first German football division. *Empirical Economics 27*, 101–112.

Dixon, M.J. and S.C. Coles (1997). Modelling association football scores and inefficiencies in the football betting market. *Applied Statistics 46*, 265–280.

Forrest, D., J. Goddard, and R. Simmons (2005). Odds setters as forecasters. *International Journal of Forecasting 21*, 551–564.

Forrest, D. and R. Simmons (2002). Outcome uncertainty and attendance demand in sport: The case of English soccer. *The Statistician 51*(1), 13–38.

Forrest, D., R. Simmons, and B. Buraimo (2005). Outcome uncertainty and the couch potato audience. *Scottish Journal of Political Economy 52*, 641–661.

Hart, R., J. Hutton, and T. Sharot (1975). A statistical analysis of association football attendance. *Journal of the Royal Statistical Society, Series C 24*, 17–27.

Kuypers, T. (1996). The beautiful game? An econometric study of why people watch English football. Discussion Paper 96-01, Department of Economics, University College London, London.

Long, J.S. and J. Freese (2003). *Regression Models for Categorical Dependent Variables using Stata*. College Station, TX: Stata Press.

[¶]Despite the prominence accorded the issue of outcome uncertainty in debate on professional team sports, clearly more than half of all the matches from Spain analyzed here (in a division with strong disparity of resources across clubs) delivered a close match, either drawn or settled by only a single goal.

Paton, D. and A. Cooke (2005). Attendance at county cricket: An economic analysis. *Journal of Sports Economics 6*, 24–45.

Peel, D.A. and D.A. Thomas (1988). Outcome uncertainty and the demand for football. *Scottish Journal of Political Economy 35*, 242–249.

Peel, D.A. and D.A. Thomas (1992). The demand for football: Some evidence on outcome uncertainty. *Empirical Economics 17*, 323–331.

Rascher, D. (1999). A test of the optimal positive production network externality in Major League Baseball. In J. Fizel, E. Gustafson, and L. Hadley (Eds.), *Sports Economics: Current Research*, Westport, CT. Praeger.

Szymanski, S. (2003). The economic design of sporting contests. *Journal of Economic Literature 41*, 1137–1187.

Welki, A. and T. Zlatoper (1999). U.S. professional football game-day attendance. *Atlantic Economic Journal 27*, 285–298.

11

The impact of post-season play-off systems on the attendance at regular season games

Chris Bojke

University of Groningen and Pharmerit UK

ABSTRACT

Post-season play-offs are featured in many professional sports leagues, and they are thought to positively influence attendance in regular season games by prolonging the extent to which teams are still in contention for end of season outcomes such as promotion. However, the variety of systems in existence indicates that the relationship between design and attendance is unknown. This research addresses this issue and aims to analyze the extent to which an example of such a play-off system influenced attendance at regular league matches during the English Division 1 2000/01 season. It does so in three steps: construction of a simple theoretical model identifying play-off relevant parameters (promotion probability; nonzero probability; significance of match); a statistical estimate of these parameter values on the effect on attendance; and finally, a predication of attendances that would be observed given the counterfactual values of promotion variables implied by a nonplay-off promotion regime.

The theoretical model identifies that play-off related variables may counteract against each other and thus makes the overall impact of play-offs on regular league attendance an empirical matter. A random effects GLM model, with a correction for endogenous variables, allows for unbiased estimates of the impact of play-off parameters when faced with strictly nonnegative, heteroscedastic and skewed data produced by heterogeneous teams. The model permits unbiased prediction of attendances under different regimes and the results show that, relative to an automatic promotion regime, the current play-off system does indeed appear to have positively influenced attendance at regular season games, though the overall impact is estimated at less than 1%. Furthermore, the redistribution of promotion probability and significance across heterogeneous teams has led to some teams benefiting more than others and raises the possibility that some teams may lose attendance during regular season games as a consequence of the addition of post-season play-offs.

11.1 INTRODUCTION

Identifying and understanding the relationship between a sporting competition's characteristics and the demand for that product in the form of attendance is an important component in the design of league and cup competition formats. One such common policy-amenable element added to league structures is the addition of a post-season play-off system, whereby the allocation of end-of-season outcomes such as winning the overall championship or promotion/relegation to different divisions is finally determined. Such play-off systems are common and featured in many diverse sports, from determining champions in North American sports such as Major League Baseball and American Football to partly determining promotion and relegation issues in open league formats such as the European soccer leagues. One motivation for the presence of play-offs is that they argued to influence regular season attendance by increasing the proportion of regular season games for which a team is still in contention for the end-of-season outcome. Though play-offs are a common feature of many professional sporting competitions, there is little consistency in the design and size of play-off structure both within and across different sports, indicating that the size and nature of the impact of play-off designs on attendance at regular season matches is largely unknown. Although there exists an extensive literature on the determinants of demand within support in general and on the impact of league design in particular, limitations in the statistical techniques and the lack of a model which relates play-off design to demand has not reduced this uncertainty (Cairns, 1990; Kuypers, 1997; Borland and Macdonald, 2003; Noll, 2003).

This chapter therefore outlines a statistical approach that may be used to address this important research gap, and is illustrated with an empirical investigation of the incremental impact of a promotional play-off system on the attendance at regular season league matches in the English professional soccer league immediately below the top-tier Premiership division. The process of conducting this research is conceptually simple and falls into three distinct steps:

1. Identification of the theoretical means by which play-off design may influence regular season games;

2. Estimation of the relevant parameters using empirical data; and finally

3. Prediction of attendances under other hypothetical play-off designs to identify the effects of different designs.

Although conceptually simple, all these steps have proved difficult in practice and methods for conducing each step are thus covered to some degree within this text, which is structured as follows: firstly, I outline a simple model of the determinants of the demand for attendance which identifies the theoretical framework by which the introduction of a post-season play-off system may influence demand in regular season games. Secondly, I describe the method by which the play-off relevant variables are derived before describing the data in section 11.4 and, in section 11.5, discussing the statistical issues which arise in estimating the parameters of the model given "awkward" skewed, heteroscedastic data generated by heterogeneous teams. Section 11.6 presents the results of the statistical estimation, finding play-off related variables statistically significant and section 11.7 predicts the incremental difference that the play-off system has made, finding a modest impact of an approximately 0.9% to 0.7% increase in aggregate attendance in regular season games in the 2000/01 season and, in addition, finding that this increase is not uniform across teams. Section 11.8 draws together the conclusions from the chapter.

11.2 THEORETICAL MODEL OF THE DEMAND FOR ATTENDANCE AND THE IMPACT OF PLAY-OFF DESIGN

The framework in which to assess the impact of play-offs on attendance is provided by a conceptual microeconomic model of demand which argues that the demand for attendance, Y_{ijt}, of a match t between opponents i and j is a function of the characteristics of that match such as: the teams/individuals competing, the cost of attendance, whether the match is televised and, importantly, the context or significance of the match in resolving who gets what end of season outcome at the end of the overall competition e.g., promotion, relegation, or the championship itself.

$$Y_{ijt} = d(x_{it}, x_{jt}, z_t), \tag{11.1}$$

with x_{it} representing home team i characteristics such as the quality of the home team, whether the team is still in contention for a desirable end of season outcome, the potential significance of the match in resolving end of season outcomes, etc. at the time of match t. An analogous set of team characteristics applicable to away team j at the time of match t is represented by x_{jt}. Finally, z_t is a set of characteristics applicable to both teams such as ticket price, whether the match is televised live, the uncertainty surrounding the outcome of the match, etc.

As attendance at matches tends to be dominated by home-team supporters one would a priori hypothesize that the factors contained within x_{it} have a larger influence than those contained in x_{jt}.

Post-season play-off systems may enter the demand function by the potential impact they have on the match significance and/or the probability of obtaining the end-of-season outcome. This may be illustrated by the example of the English soccer leagues where prior to the introduction of play-off systems, the second highest division operated a strictly automatic promotion scheme whereby the teams which finished in the top three positions at the end of the season were automatically promoted to the higher division. In contrast, the current play-off design is one in which the top two teams get automatically promoted to the higher division and the following four teams (positions three through six) play in a cup-style knock-out competition in which the winner joins the two automatically promoted teams. Given those definitions, imagine a hypothetical two-thirds played season where a mid-table team no longer has any reasonable chance of obtaining third position but a distinct possibility of obtaining sixth or slightly higher, under the automatic promotion regime, this team has a zero probability of obtaining promotion, whereas under the stated play-off system, there would be a nonzero probability — thus illustrating the potential impact of a post-season play-off function on the determinants of demand during regular season games.

Indeed this impact appears to be singled out as a main motivating factor for the presence of play-off systems: a nonzero probability of obtaining promotion is a positive demand driver; play-off systems create more games where teams have such nonzero probabilities, ergo play-off systems increase demand in regular season games. However, this simple conclusion may be erroneous as it omits several other important issues that may act in a counteractive manner: for example, what is the impact of such systems on the significance of a given match in determining end-of-season outcome? If there are no counteractive forces then an attendance maximizing play-off design would appear to include all competing teams. Indeed this is almost the case in the US Major League Soccer (MLS) where eight out of ten teams qualify for the end-of-season championship play-offs, prompting former US national coach Bruce Arena to comment that "most of the MLS regular season games mean nothing" (Gardner, 2005).

A further issue to consider is that with a fixed number of promoted teams then at any point in time the probability of promotion for a given team summed across all teams is equal to the number of promotion spots. That is, if there are three promotion places, then if at a given time point we were able to measure the probability of an end-of-season promotion for each and every team, they would all sum to 3. Thus when a play-off system creates more games with nonzero probabilities, it does so by redistributing the probability of promotion across teams and games rather than increasing probability in total. As there exist heterogeneous football clubs addressing very different sized markets, redistributing promotion probability, particularly from bigger clubs to smaller clubs as is likely in a play-off system, may potentially reduce attendance.

The above discussion has identified three demand inputs that play-off systems may affect: (i) the probability of obtaining the outcome (promotion in this case) for team i at match t, p_{it}; (ii) a simple dummy variable indicating whether this probability is nonzero or not, nzp_{it}; and (iii) a significance variable, in this case defined as the difference in the probabilities between a win and a defeat for team i in the match t would make, dp_{it}. However, identification (and measurement) of these variables is insufficient to estimate the impact of different play-off designs, in addition one must identify the relationship between these variables and play-off design.

In order to do this it is useful to consider the probability of promotion as the product of two different probabilities: the probability of team i finishing in league position m, $\pi_{it}(m)$ and the probability of a team finishing in position m being promoted, $\pi(prom|m)$, where the lack of subscripts on this latter term indicates that it is constant across all teams and invariant to the stage of the season.

Thus, in terms of our demand variables:

$$p_{it} = \pi_{it}(m) \times \pi(prom|m), \tag{11.2}$$

$$nzp_{it} = \begin{cases} 1 & p_{it} > 0, \\ 0 & p_{it} = 0, \end{cases} \tag{11.3}$$

$$dp_{it} = p_{it+1}(i \text{ wins } t) - p_{it+1}(i \text{ loses } t). \tag{11.4}$$

This is easily illustrated by comparing the English system pre- and current play-off systems. In the old automatic scheme:

$$\pi(prom|m) = \begin{cases} 1 & \text{if } m = 1, 2, 3, \\ 0 & \text{otherwise} \end{cases} \tag{11.5}$$

Whereas in the current play-off scheme it would appear to more like:

$$\pi(prom|m) = \begin{cases} 1 & \text{if } m = 1, 2, \\ 0.25 & \text{if } m = 3, 4, 5, 6, \\ 0 & \text{otherwise.} \end{cases} \qquad (11.6)$$

In the automatic scheme these probabilities are known with certainty, whereas in the current play-off scheme they are estimated with some uncertainty. The assumption of equal probabilities of promotion from each of the four qualifying positions is supported by results from the second tier of English soccer but may not apply in other circumstances and so further research in this area may be required (Dart and Gross, 2006). Nevertheless Equation (11.2) through Equation (11.6) identifies the potential impact play-off systems may have on attendance on regular league matches and can be used to evaluate claims of the impact of play-off design on attendance.

The theoretical model has thus identified a potential trade-off, marginal fans of clubs may be attracted to matches in which their team still has a nonzero probability of obtaining a desirable end-of-season outcome and play-off systems can increase this number of games. However, since the overall probability is fixed, this effect is achieved by a redistribution of probability. Therefore, the impact of a play-off system may not be theoretically determined, but is instead an empirical question of which of these counteracting forces dominates.

11.3 MEASURING THE PROBABILITY OF END-OF-SEASON OUT-COMES AND GAME SIGNIFICANCE

The play-off related variables may only be implemented in a regression model if there are measures of these probabilities readily available. Though there exists a betting market for end-of-season outcomes potentially providing a means of obtaining probabilities of end-of-season outcomes for each team prior to each match (and by definition asserting whether the probability is nonzero),[*] there remains one element which is not readily available: how a match may affect probabilities of obtaining end-of-season outcomes. In practice this is more complex than simply including market betting odds of promotion or winning the championship at the time of the match (which may be observable) as one must include the betting odds should that match be won or lost, all other things remaining equal (potentially neither of these odds will be observable.) Thus even if the conceptual measures are accepted, estimation of the difference in probabilities is likely to remain a contentious area in research.

To address this issue I implement an imperfect measure based on placing simulated results for remaining matches (based on individual match ex-ante betting odds) onto the existing league table at any point in time. For example, suppose there are 40 matches remaining in a 500 match league. The 460 completed matches give a factual table; where each team is, how many points they have, how many games they have

---

[*]Although these data are not available for this research.

played, who has played who, etc. The probabilities of the game outcomes (home win, draw, away win) for all 40 remaining games then allow us to devise a measure of the likely probabilities of where each team will finish. That is, I can simulate the outcome of each remaining game and add them to the fixed real table to produce an expected final table. Conducting this simulation exercise a number of times allows probabilistic statements about the likely finishing positions, i.e., for each team we can simulate $\pi_{it}(m)$ for each m and obtain p_{it} by multiplying each $\pi_{it}(m)$ by the assumed values of $\pi(prom|m)$ for each m.

Furthermore, one can obtain a measure of the significance of any of the remaining matches by: first taking the fixed existing league table, assuming team i wins that game and simulating the remaining games and recording the probability of finishing positions for team i, then, secondly conducting the same exercise with the exception that we assume team i loses that particular game. The differences in these two probabilities of finishing in a particular league position, multiplied by the relevant $\pi(prom|m)$ and summed across all m therefore gives a measure of the significance of that game, i.e., the difference in the probability of a team obtaining the end of season outcome as a result of winning the game and losing that game.

The proposed measures have the benefit of incorporating all teams' actual positions and points and the expected total number of points allowing for the market expectations of the difficulty of the remaining schedules, taking into account matches between teams effectively competing for the same places and the structure of the league, i.e., how many automatic promotion places there are, etc. However, the measures do have a number of weaknesses. Firstly, data on betting odds are only available for the odds immediately prior to matches being played. Ideally, contemporary betting odds for all matches at each time point during the season are required in order to calculate the true final table expectations. If odds change over time, then by using odds posted in March to estimate the expected significance of a match in January, we are likely to have measurement error in a right-hand side variable. If odds are subject to random changes over time, then this will simply be captured in the error term and will not cause any specific problems. However, if the odds are subject to systematic changes, a more likely scenario, then some bias via measurement error will occur.

Secondly, I assume that the outcome of a particular match has no influence on eventual league outcomes other than that made by the allocation of points from that game. In other words I assume that there are no spill-over effects; that by winning one game, a team does not influence the probabilities in another match involving itself or other teams.

Thus the measures I use to provide values for the model are potentially flawed, and flawed in a systematic and predictable manner. However, the approach incorporates in a systematic manner: the reality of the existing tables, the structure of the league and a means of incorporating expectations based on an existing schedule (i.e., who plays who with a probabilistic statement of the likely outcomes). In addition, the empirical results have face validity in that they identify games as significant which look to the author as games which are worthy of that description (see section 11.4). Nevertheless, a more rigorous solution to measuring these probabilities is likely to be a future research priority in this area.

11.4   THE DATA: THE 2000/01 ENGLISH FOOTBALL LEAGUE SECOND TIER

Promotion (and relegation) play-offs were introduced to the English football leagues in the 1986/87 season. The current system in England for the division below the Premiership, now known as the Championship, but during the 2000/01 season as the 1st Division, is that the top two teams are promoted automatically to a higher division and the next four teams enter a straightforward knock-out tournament. The 3rd and 4th placed teams meet the 6th and 5th placed teams respectively in a two-legged semi-final, playing home and away. The winners of the two semi-finals then meet for a single final played on neutral territory where the winner is promoted to the higher division. Currently, in all divisions, relegation is strictly automatic and based on league position. The nonplay-off alternative in the English system is that the top three placed teams gain automatic promotion.

Over the course of a regular season each of the 24 teams plays a balanced schedule, with each team playing each other twice, once home and once away and so a full 1st division season consists of 552 games. For each league match, three points are awarded to a winning team and none to a losing team. In the event of a draw, a single point is awarded to each team. League positions are firstly determined by accumulated points, then aggregate goal difference and then total goals scored if teams are tied on points. The season runs from August to May with no winter break. Fixtures are determined prior to the start of the season, but have some degree of flexibility with some game dates being rearranged due to weather postponements, cup matches, or television schedules. Typically, rearranged fixtures will occur as close to the original date as possible, for example, games moved so they may be televised live are typically moved from a Saturday to the preceding Friday, following Sunday, Monday, or Tuesday. Fixtures are traditionally played on Saturday afternoons and also on bank holidays such as Boxing Day and Easter Saturday and Monday.

During the 2000/2001 season, the teams within the 1st division showed sizeable intra- and inter-team variation in the match attendance at their 23 home games, as shown in Table 11.1.

The table shows a number of notable features: there is a wide variation in average attendances with larger clubs such as Nottingham Forest, Birmingham, Blackburn, Sheffield Wednesday, and Wolverhampton Wanderers having averages three to four times that of the smaller clubs Grimsby, Crewe, Stockport, and Wimbledon. Indeed none of the minimum attendances at the four biggest clubs fall below the maximum attendance at the smaller clubs. Clubs with the largest average values also tended to have the largest variances, with the variance being correlated with the square of the mean, indicating heteroscedasticity. That the means are systematically closer to the minimum level rather than the maximum is indicative of a skew. The data also indicate that stadium sizes are not thought to have had a censoring effect on the data, with only a very few attendances approaching stadium capacities.

The teams with larger and smaller average attendances are predictable on historical grounds, may reflect different potential team specific market sizes, and generally afford the teams greater or fewer resources which can aid league performance.

TABLE 11.1

2000/01 Attendances, league positions, and simulated promotion
probabilities by team.

Team	Mean Att.	St. dev. Att.	Median. pos.	Mean Prob.	Mean Prob. Dif	Nonzero Prob. Games
Barnsley	14465	1928	13	0.012	0.007	27
Birmingham	21283	3798	4	0.324	0.093	46
Blackburn	20740	3544	4	0.541	0.125	46
Bolton	16062	3577	3	0.412	0.113	46
Burnley	16234	1872	9	0.052	0.035	45
Crewe	6698	983	18	0.001	0.001	8
C Palace	17061	1986	19	0.003	0.003	15
Fulham	14990	2734	1	0.977	0.015	46
Gillingham	9281	735	15	0.002	0.002	23
Grimsby	5646	1212	19	0	0	3
Huddersfield	12809	2673	23	0	0	5
Norwich	16525	1835	16	0.002	0.001	24
N Forest	20615	2456	7	0.112	0.06	45
Portsmouth	13533	1973	15	0.001	0.002	15
Preston NE	14617	1411	6	0.118	0.053	46
QPR	12013	2112	22	0	0	5
Shef United	17211	4313	9	0.039	0.029	42
Shef Wed	19268	5272	21	0.001	0.001	6
Stockport	7030	1300	20	0	0.001	4
Tranmere	9052	1255	19	0	0	6
Watford	13941	1939	5	0.234	0.072	44
WBA	17657	2097	5	0.157	0.066	46
Wimbledon	7901	2343	11	0.029	0.024	44
Wolves	19258	3067	15	0.002	0.002	16

Hence the larger clubs tend to be found at the higher end of the league table with a few exceptions: Sheffield Wednesday, a team with a historically large support, had relatively few resources as a result of overspending in previous seasons. Sheffield Wednesday spent much of the season in the bottom four positions, but still managed an impressive home average of 19268 attendees. Conversely Fulham, a team with a rather more modest historical following, was a team that benefited from a wealthy chairman and spent almost all of the season in first place. Despite the league success, crowds failed to match those of Sheffield Wednesday.

Such summary statistics hint at the importance of allowing for individual team specific latent support that is independent of short term success (or other temporary variables). Indeed such data hint at the potential limitations that tinkering with league design may have on attendance.

With regard to the measure of promotional significance, if we assume that the probability of promotion when qualifying for a play-off position is 0.25, then the average league position, probabilities of promotion, the difference the game may make to the probability of promotion facing each team immediately before their next match, and the number of nonzero probability of promotion games are shown in the final four columns. Fulham dominated the division, occupying the top spot for most of the season. With a high proportion of games won, a sizable gap between Fulham and the chasing group developed, and taking the likelihood of winning their games during the remaining schedule into account, the probability of being promoted was close to one at the time of many of their matches (though this is probably an overestimate for the earlier games created by the means of generating the probabilities). The difference an individual match would make in altering the probability of promotion for Fulham was small, averaging around 1.5% per game.

At the opposite extreme, there are a number of teams whose probabilities are so low (15 out of 24 teams have average probabilities of promotion of less than 5%) that, similar to Fulham, the average difference a game may make to the probabilities of promotion is small (less than 3% for the 15 teams). However, for those teams with an average probability of promotion in the range 0.25 to 0.75, there is an average probability difference per game of approximately 10%.

Figure 11.1 shows the relationship between the probability of promotion and the difference to probabilities that a match can make. It shows a quadratic relationship between the two with low and high probabilities being associated with low differences in potential changes, with a few obvious outliers. The limited nature of the observational data and the mathematical constraints imposed by probabilities may limit the ability of the statistical methods in separating out the effects of probability and the effect of the difference in probabilities a game can make.

Given the flawed nature of assessing the probabilities, a note is required to support the practical means of measuring the probability of promotion and the difference in probability a match may make. This is illustrated by inspection of games for which the model predicts a high significance. The most significant match is thus a match in February. Initially, one might feel that this would be too early in a season which runs to May; but closer inspection of the game involved is fairly convincing. The game in question occurred when second-placed Bolton hosted third-placed Blackburn. At

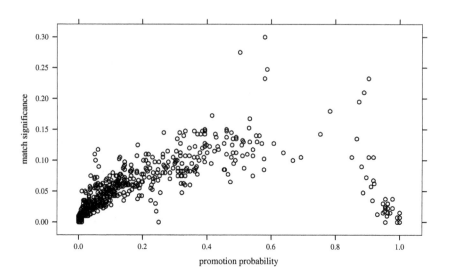

FIGURE 11.1

The relationship between promotion probability and match significance.

the time Bolton was seven points clear of Blackburn (66 points against 59, Fulham in first place had 75 points, Birmingham in fourth place also had 59 points), and had thirteen games left to play in the regular season. Blackburn had played one game less and had fourteen matches left. The probabilities of promotion prior to the match were: Bolton 58% and Blackburn 50%. If Bolton were to win the match, their lead over Blackburn would increase to ten points and the probability of promotion would increase to 76.75% and Blackburn's would fall to 34.75%. If Bolton were to lose, their probability would fall to 46.75% (hence, a probability difference of the game of 30%) and Blackburn's would rise to 62.25% (hence, a probability difference to Blackburn of 27.5%[†]). The measure has thus incorporated the distance of the challenging teams, the fact that two of the challenging teams were playing against each other, that the third-placed team had a game in hand and the importance of finishing second over third, etc.

Blackburn had a similarly significant match two games later when they played away at Birmingham City. With both teams having a probability of promotion of approximately 59%, the difference to Blackburn of the game was 24.75%. The slightly smaller difference in probability reflects the fact that this match was not expected to affect the points total of the current second-placed team which was still Bolton; ahead of third-placed Birmingham on goal difference and three points ahead of Blackburn.

[†]The final score was Bolton-Blackburn 1-4.

TABLE 11.2
Impact of regimes on aggregate significance.

	Current Play-Off System	Automatic Promotion System
Mean promotion probability (home)	0.128	0.129
Mean match significance (home)	0.029	0.03
Nonzero probability games (home)	330	216
Mean promotion probability (away)	0.124	0.123
Mean match significance (away)	0.03	0.029
Nonzero probability games (away)	323	215

Of the remaining games, all of the significances over 15% feature either Birmingham, Bolton, or Blackburn, mostly in the later half of the season and mostly reflecting the chase for second place. None of the games that look like being major influences on play-off places at the bottom end of the qualifying places (i.e., positions 5 and 6) are valued quite as significantly. The clearest example appears to be Nottingham Forest away to Wimbledon in their third to last game of the season, where seventh-placed Forest was two points adrift of the last play-off position and faced a likely promotion probability difference of 11.75% (14% if Forest won and 2.25% if they lost, Wimbledon was not a play-off contender). Although the probability of promotion and difference in probability is relatively modest compared to some games, since Forest was ten points adrift of the third place at the time, with only a maximum of nine points available, if it were not for the play-off regime, the probability of promotion and the difference that match could make to the probability of promotion would be 0%.

The proposed measure of significance has identified games which appear intuitively significant in the context of the season and clearly incorporate notions of six-pointers (i.e., games where points could be denied to other competing teams), actual position, and expected performance over the remaining season. Arguably such benefits provide a powerful motivation for persisting with imperfect measures.

Finally, given an understanding of the relationship between play-off systems and promotion probabilities and game significance and assuming that match results are unaffected by the league design, then it is possible to estimate the probabilities and match significances for this season as if it had operated under the traditional pre-1987 league structure with the teams finishing in the top three league positions being automatically promoted to the top division.

Table 11.2 shows the difference in aggregate levels of probability and significance under the two different regimes: the current play-off system and the pre-1987 automatic promotion system for home and away teams.

The table shows that the average probability of promotion per game is, as expected, unchanged (there is a slight redistribution between home teams and away teams which is not inconsistent with the mathematics). The situation is the same for the difference in probabilities a match may cause; the aggregate stock of this measure is unchanged across regimes. This may initially seem counter-intuitive, but the limited nature of the play-off system described has essentially moved the significance from games which distinguish finishing third from a lower position to those that determine who finishes second from a lower position.

The obvious and expected difference is in the number of games for which either the home team or away team has a nonzero probability of promotion. In the 2000/01 season, the current play-off system created 114 more games where a home team had a nonzero probability than would occur with an automatic promotion scheme, approximately 21% of the season.

11.5   STATISTICAL ISSUES IN THE MEASUREMENT OF THE DE-TERMINANTS OF ATTENDANCE

As indicated by the descriptive analysis, a number of statistical issues in the estimation of the relevant parameter values of the demand model exist: attendance data are nonnegative, heteroscedastic, and often skewed. In addition, attendances are clustered within heterogeneous teams. Finally, conceptually different important variables such as dp_{it} and p_{it} may be highly correlated causing issues of multicollinearity.

11.5.1   SKEWED, NON-NEGATIVE HETEROSCEDASTIC DATA

Within the economic literature, the most commonly addressed data problem is that of a positively skewed dependent variable and heteroscedasticity. This tends to be solved via a semi-log or log-log functional form (Forrest and Simmons, 2002; Garcia and Rodriguez, 2002), i.e., defining the dependent variable as the natural logarithm of attendance and computation of robust standard errors for valid inference. Such an approach is acceptable if the objective of the analysis is to conduct hypothesis tests. However, the objective of this research is not only to conduct a hypothesis test on the statistical significance, but also to identify the incremental effect of the system measured in its natural metric, i.e., attendances. In this case the limitations of such transformation methods are revealed as: "unbiased and consistent quantities on the transformed scale usually do not retransform into unbiased or consistent quantities on the untransformed scale" (Duan, 1983). Although Duan also identified a means of correcting this bias, smearing, it has been found that the smearing procedure performs poorly in the presence of heteroscedastic error terms (Manning and Mullahy, 2001).

The alternative then is to look for models which can accommodate different data assumptions and thus circumnavigate the retransformation problem altogether. The generalized linear model framework (GLM) represents a class of models (including the familiar linear model, logit, probit, and Poisson count models) which provide some scope in handling the skewed heteroscedastic nonnegative attendance data.

McCullagh and Nelder (1989) provide a full description of the GLM framework, but a brief summary follows. In a simple linear model a relationship between a set of explanatory variables, x_i, and a response variable, y_i, is typically modelled as $y_i = v_i + \epsilon_i$. Where v_i, the linear predictor, is given by $v_i = x_i'\beta$ and, the error term, the random fluctuations around the outcome variable are assumed to be given by $\epsilon_i \sim \mathcal{N}(0, \sigma^2)$.

The GLM specification introduces two additional elements. First, it splits the relationship between the linear predictor and the conditional expectation, such that the expectation is a function of the predictor, i.e., $\mu_i = g^{-1}(v_i)$ or $g(\mu_i) = v_i$, such that $g(E(y_i|v_i)) = v_i$. The function $g(\cdot)$ is known as the link function as it provides the link between the linear predictor and the conditional expectation. In the simple linear model, as $\mu_i = v_i$, the link function is simply known as the identity link function. However in the GLM framework other link functions may be applied such as a log or probit link function, i.e., $\log(\mu_i) = v_i$ or $\Phi^{-1}(\mu_i) = v_i$. The inverse of the link function, $g^{-1}(\cdot)$ may be used to turn a given linear predictor into an expectation on the original metric e.g., $\mu_i = \exp(v_i)$ or $\mu_i = \Phi(v_i)$.

Though in principle this looks very similar to the transformation solutions, there is a subtle but important distinction: GLM deals with a function of the expectation, $g(E(y_i|v_i)) = v_i$, while the transformation solutions specify the expectation of a function $E(g(y_i)|v_i) = v_i$. It is this distinction which means the transformation solution is plagued by the back or retransformation problem and the GLM approach is not.

The second distinguishing feature of the GLM is the increased range of distributions which surround the conditional expectation, i.e., the conditional probability distribution of the responses. For example, in the linear model, the conditional distribution is given by $\mathcal{N}(0, \sigma^2)$. In the GLM framework, the conditional distribution may be any from the exponential family — this includes the binomial, normal, and Gamma distributions. The choice of conditional distribution implies a relationship between the conditional variation and the conditional mean given by:

$$\text{var}(y_i|v_i) = \phi V(\mu_i), \qquad (11.7)$$

where $V(\mu_i)$ is known as the variance function and links the conditional variance and expectation, and ϕ is known as the dispersion parameter which is unrelated to the condition expectation. For example, with the Poisson distribution, the variance is equal to the expectation (a well known property of the Poisson count model); and with the Gamma distribution, the variance is proportional to the square of the expectation. The observed empirical relationship between variance and conditional mean can therefore be used to determine which is the appropriate choice of distribution. In this case, given the summary data, the Gamma distribution appears to be an appropriate choice to model the nonzero, skewed, and heteroscedastic attendance data.

The other modelling choice is to choose an appropriate link function. A log link function creates an multiplicative effect between covariates. In the next section I discuss the addition of individual team-specific effects to account for clustering, with a log link function this team specific effect interacts with other covariates, thus allowing some scope in modelling a heterogeneous response to play-off related covariates across teams (a similar motivation is provided by Berri, Schmidt, and Brook, 2004, in a log-log specification.)

Thus in order to produce unbiased estimates of the incremental effect on attendance of the English play-off system relative to an automatic promotion scheme, given the skewed, heteroscedastic, and nonzero nature of the attendance data, a GLM model with a log link function and Gamma distribution is specified.

11.5.2 CLUSTERING OF ATTENDANCE WITHIN TEAMS AND UNOBSERVED HETEROGENEITY

Like the simple linear regression models, the standard GLM framework assumes that the observations are independent of each other. However, the descriptive analysis has indicated that unobserved heterogeneity may exist between teams, principally in this case different underlying market sizes, which in a classical multi-level setting may be captured by a random intercept or random effect consistent across time, v_i.

Conceptually, the random term is easily incorporated in the GLM framework in the linear component, i.e., $v_{ijt} = x'_{it}\beta_i + x'_{jt}\beta_j + z'_t\beta_t + v_i$ and further extensions could be made to allow for away team random effects. However, a key assumption of the random effects model is that the random effect is assumed to have a zero correlation with the included observed variables. In this case this assumption appears suspect as we may expect the teams with larger underlying markets to have access to larger resources and hence have higher expected probabilities of promotion, etc. The consequences of ignoring this potential correlation are that we may obtain biased estimates of the impact of the correlated covariates.

Skrondal and Rabe-Hesketh (2004) describe these correlated observed covariates as endogenous and identify that many analysts (particularly economists) falsely believe that this situation automatically rules out a random effects model in favor of a fixed effects model. This is not the case as a rather simple and elegant solution exists: the impact of a correlated covariate x_{it} may be estimated without bias in a random effects model if one simply includes the cluster mean $\bar{x}_{.t}$ as an additional covariate in the regression model. The inclusion of the cluster mean breaks the correlation between the random effect and covariate of interest.

Theoretical considerations guide the identification of potentially endogenous variables. For example, the three play-off related variables for home teams are all likely to be endogenous and so team clustered means are included for these variables in the regression model. However, variables associated with away teams (such as away team league position) are likely to be exogenous as by the random nature of fixture determination, the values of the away team variables will be uncorrelated with the home team random effect. Other potential endogenous variables include whether the match is broadcast live on the (Sky) satellite subscription channels and ticket prices.

11.5.3 MULTICOLLINEARITY

A potential estimation problem occurs through the collinear nature of several of the explanatory variables. Predictably those variables associated with the probability of promotion are all highly and significantly correlated. For example, the correlation between a home team's league position and the significance of the game to a team is -0.69. The consequence of such collinearity is that we may be unable separate out the individual effects of variables and hypothesis tests may have large (but still unbiased) standard errors (Kennedy, 2004). Further consequences may include parameter estimates having incorrect signs or implausible magnitudes and that parameter estimates are sensitive to inclusion (or exclusion) of a few data points (Greene, 2003). Unfortunately, to some extent we are constrained by the passive nature of the data collection — we are unable to construct an active data collection via an experimental design whereby we can produce orthogonal relationships between variables of interest. It is thus worthwhile exploring potential solutions to the problem.

An obvious solution may be to incorporate more data by expanding the data to include more years or more divisions. However, if the relationship between these variables holds across other possible datasets, then this solution will be of limited use — the extra data will contain little additional information, though the increased sample sizes would, at the margin, reduce parameter estimate variance. That additional data would contain the same restricting correlations is very likely: it is hard to imagine leagues where league position is not highly correlated with the probability of obtaining an end of season outcome and where low (or high) probabilities of obtaining that outcome are subject to large swings as a result of one game.

Other analytical solutions such as ridge regression or factor analysis are often implemented but also criticized (Maddala, 2001), and it is likely that there exists no analytical solution to multicollinearity in this application. For example, collapsing the three play-off related variables to a single factor may resolve the issue of collinearity but does not permit estimation of the separate effects. In such circumstances it may be more prudent to accept the limitations of the data, to concede that the presence of multicollinearity is unavoidable, and to examine the sensitivity of the results and policy implications to the problems/uncertainty caused by multicollinearity. In this particular case, I do this by presenting a range of models where variables whose non-significance or magnitude is considered a potential artefact of multicollinearity are variously dropped and the sensitivity of the results to each specification examined. The intuition behind this approach is that it gives us the range of responses from models with different assumptions. If we find that predictions are largely invariant to model specification, then the consequences of multicollinearity are not severe.

11.5.4 FINAL STATISTICAL MODEL

All final statistical models are thus of the form given in equation 11.8 expressed as a generalized linear model:

$$\mu_{ijt}^{po} \equiv E(y|v_{ijt}^{po}) = g^{-1}(v_{ijt}^{po}), \tag{11.8}$$

i.e., the conditional expectation of attendance between home team i against away team j at a league match indexed by t under the play-off system (po), μ_{ijt}^{po}, is a function $g^{-1}(\cdot)$ of a linear predictor v_{ijt}^{po}. Where $g^{-1}(\cdot) = \exp(\cdot)$, and $v_{ijt} = x_{it}^{po}\beta_i + x_{jt}^{po}\beta_j + z_t\beta_z + \bar{x}_{.t}^{po}\beta_{end} + v_i$.

x_{it}^{po}, x_{jt}^{po}, and z_t are vectors of match characteristics relating to the home team, away team, and match at match t. The lack of a play-off superscript for the match specific characteristics indicates they are considered invariant to play-off design. v_i is a team specific time invariant random effect ($v_i \sim \mathcal{N}(0, \sigma_i^2)$), $\bar{x}_{.t}^{po}$ is a vector of team averages of those variables which are potentially endogenous and β_i, β_j, β_z, and β_{end} are vectors of unknown parameters to be estimated. Note that the component $x^{po}_{.t}\beta_{end} + v_i$ accounts for the underlying heterogeneity of each team split into a fixed and random component and does not represent an impact of any observable match characteristic.

The data indicate that the variance of the dependent variable is related to the square of the expectation, a Gamma distribution is assumed and hence the conditional variation is given by:

$$var(y_{ijt}|v_{ijt}) = \mu^2_{ijt}\alpha^{-1}. \tag{11.9}$$

11.6   MODEL ESTIMATION

11.6.1   CHOICE OF EXPLANATORY VARIABLES

The GLM statistical model with a log link and Gamma distribution estimated with a random effect for the home team has been justified in sections 11.4 through 11.5. Here I outline the inclusion criteria for the explanatory variables.

Included are the three play-off related variables: p_{it}, nzp_{it}, and dp_{it} for the home team and a set of analogous variables for the away team p_{jt}, nzp_{jt}, and dp_{jt}. Team quality is captured by league positions for both home and away teams: pos_{it} and pos_{jt}. Matches against local rivals are also anticipated to generate interest. Variables for derby games for both home teams and away teams ($derby_{it}$ and $derby_{jt}$) are constructed on the basis of nearest neighbor and are set to zero unless the away (or home) team featured in that match is the nearest club to that particular team, where it is set to one. The one exception to this construction rule is the intra-city Sheffield derby games: due to the intense rivalry between Sheffield Wednesday and Sheffield United, a further dummy variable is constructed, $sheffd_t$, to capture the difference between these two intra-city derby matches and any other derby games. $miles_t$ measures the distance between teams as attendance is expected to diminish in distance as travelling away fans may be deterred by the additional travel costs and marginal home fans may be deterred by a lack of interest in seeing a team from some distance. Two further dummy variables, $first_{it}$ and $last_{it}$, are constructed for the home team only, to capture whether it was the first or last home game of the season for that team. Clubs often arrange additional entertainment and teams may complete a post-match lap of honor to say thank you for the fans' support throughout the season. The additional party atmosphere of a home team's first or last game and the first (or last) chance to see your team at home since last (until next) season is expected to attract the marginal fan. The same arrangements are not made for away teams and thus no variable for the away team is included.

Match uncertainty has been the focus of several empirical studies and is thought to be a driver of demand — with fans expected to shy away from games which have a certain outcome. In this chapter I measure the outcome uncertainty as a Theil measure based on the match betting odds, (Peel and Thomas, 1996). The measure is increasing in uncertainty and ranges from 0.75 (Fulham versus Tranmere, where the bookmakers average mark-up purged probability of a home win was 0.73; a draw was

0.18; and away win was 0.09) to 1.09 (Crystal Palace versus QPR, where probability of a home win was 0.37; a draw was 0.29; and away win was 0.34). Data on betting odds, as with data on fixtures, results and game dates, was provided by Mabel's Tables (2003) and data on attendance provided by Statmail (2003).

Football is traditionally played on a Saturday afternoon and on certain public holidays. However, many fixtures are played midweek. It is perceived that, relative to the traditional Saturday fixture, games played on public holidays attract attendance whereas games played midweek deter attendance. Thus dummy variables $midweek_t$ and hol_t are included, capturing when the game was played (the omitted baseline category is a Friday or weekend fixture). Games played on a weekday bank holiday, such as a Bank Holiday Monday, are classed as holiday fixtures and not midweek fixtures. A sky_t dummy is included, capturing whether the game was televised live on the main subscription satellite TV provider, Sky. Dummy variables Aug_t, Sep_t, Oct_t, Nov_t, Dec_t, Feb_t, Mar_t, Apr_t, and May_t capture the month in which the game is played with January being the omitted category.

However, a number of desirable variables are omitted on the grounds that they were unavailable. The home team random effect is argued to pick up the net impact of time invariant omitted variables. However, with variables such as ticket price, which for some teams may be responsive to expected demand for individual matches and therefore vary within a season, a random effect may not suffice. Common examples include children under 16 being admitted for £1 when accompanied by a full-price adult or season ticket holders being able to bring a friend for £5. The omission of this endogenous variable means the model must be regarded as a reduced form model rather than fully structural. Though if ticket prices remain constant within teams across a season then the effect of ticket price will be subsumed in the random effect.

Finally, given the requirements of the random effects model, a distinction must be made between those variables that are regarded as endogenous and those that are exogenous. For those variables that are considered endogenous, team season means are included in the model specification. All those variables associated with the away team are assumed exogenous as are holiday fixtures, derby variables, miles measure, month dummies, and first and last game variables. The play-off related variables for the home team are considered endogenous as are the following: the home team's league position, the Theil measure of match uncertainty, and whether the match is televised live on Sky. In addition since televized games are often moved to midweek, then the midweek dummy is also considered potentially endogenous.

11.6.2 REGRESSION RESULTS

In order to test the sensitivity of the analysis to issues of collinearity between league positions and play-off related variables, five models are estimated. Model 1 has all explanatory variables included; models 2 through 5 omit league positions. Model 2 has all the play-off related variables included whereas models 3, 4, and 5 have a single play-off variable included in each case: nonzero probability dummy, promotion probability, and probability difference (or significance) respectively. All regressions are estimated using the GLLAMM suite of commands (Rabe-Hesketh, Skrondal, and

TABLE 11.3
Regression results model 1.

Variable	coeff.	std. err.	Variable	coeff.	std. err.
Constant	23.51	0.855	September	−0.049	0.03
pos (home)	−0.004	0.002	October	−0.004	0.028
pos (away)	0.002	0.002	November	0.007	0.03
nzp (home)	0.058	0.022	December	0.051	0.028
p (home)	0.302	0.101	February	0.09	0.029
pd (home)	0.359	0.336	March	0.107	0.029
nzp (away)	0.003	0.019	April	0.092	0.028
p (away)	0.194	0.034	May	0.075	0.05
pd (away)	−0.247	0.191	av.pos (home)	0.034	0.006
hol	0.052	0.02	av.nzp (home)	0.361	0.073
first	0.112	0.039	av.p (home)	−2.504	0.176
last	0.108	0.037	av.pd (home)	1.419	0.422
midweek	−0.067	0.016	av.sky	1.092	0.129
derby (home)	0.06	0.035	av.real mi k	−0.556	0.141
derby (away)	0.05	0.035	av.theil	−14.595	0.818
sky	−0.067	0.023			
theil	0.529	0.165	var (rand effect)	0.033	0.001
miles	−0.001	0			
sheffd	0.507	0.095	log likelihood	−4968.423	
August	−0.061	0.037			

Pickles, 2004) in Stata SE 8 (StataCorp, 2003) and use a random effects specification with a log link function and a Gamma distribution. There are 552 observations clustered within 24 home teams.

Table 11.3 contains the regression result and regression log-likelihood for model 1 and the results for the remaining models are available on the website associated with this book: www.statistical-thinking-in-sport.com. The results are broadly similar across models and any meaningful differences are discussed within this text.

The results in all models conform to prior expectations: the play-off related variables are all positive and in most cases statistically significant; parameters associated with the home team exceed their analogous away team counterparts; variables anticipated as being positive drivers of attendance such as games being played on bank holidays, first and last games of the season, and derby games are all positive and mostly significant; variables anticipated to have a negative impact on attendance such as the live broadcast variable, increasing distance between teams, and a midweek setting all have negative coefficients estimated which are mostly statistically significant. In accordance with economic theory, the positive and mostly significant variables estimated for the Theil measure of uncertainty support the notion that increased uncertainty has a positive impact on demand.

The month dummies show a distinctive time trend with months after January having positive coefficients whereas months prior to January having negative coefficients. If the month dummies were picking up a negative influence of adverse weather conditions, then one would have expected the months of August and September to have had positive coefficients. Thus, the month dummies may be picking up some omitted variable which is correlated with the time of the season. Potentially, this may be an artefact of the means of calculating the promotion probabilities and match significances. Matches earlier in the season have promotion probabilities and significances calculated using a disproportionate amount of matches simulated on the basis of betting odds fixed some months later, thus these variables in the earlier part of the season may be more prone to measurement error — the trend observed in the month dummies may be an indication of this and indeed a crude means of correcting this bias.

The averaged endogenous variables are generally significant and with the expected signs. These variables do not estimate the impact of the associated variable but contribute to a fixed component of a team's unobserved heterogeneity. For example, the positive and significant coefficients associated with the averaged Sky variable demonstrate the tendency for games with expected high attendances to be chosen for live broadcast rather than a positive impact of broadcasting on attendance. The negative and significant coefficient estimated for the Sky dummy is the unbiased estimate of the impact of live broadcast, indicating a negative impact on attendance.

In all cases the random effects specification appears justified with the estimated variance of the random effect being significantly different from zero.

The variation in results across models is also as expected. Those variables uncorrelated with the play-off variables produce estimates generally consistent across the models as various play-off variables are omitted. As expected, the coefficients associated with play-off related variables do change as correlated variables are omitted. Notably the magnitude and significance of the remaining play-off variables increases as correlated variables are omitted.

The promotion probability variable is positive and significant in all models for both home and away teams. The nonzero probability dummy is also always positive and significant for the home team, though only significant for the away team when no other play-off related variable is included. The probability difference or significance variable is of the expected sign but only significant when it is the only included play-off related variable and even then, only when it is applied to the home team.

11.7   THE IMPACT OF THE PLAY-OFF SYSTEM ON REGULAR LEAGUE ATTENDANCES

Given a theoretical model, a measure of the expected values of the play-off related variables under an automatic promotion regime and estimates of the relationship between these variables and attendance, it is possible to predict the expected attendances for each match that would have occurred had the play-off system not been implemented by replacing the observed play-off variable values with the estimated counterfactuals, as shown in equation (11.10). As previously stated this requires the

TABLE 11.4

Incremental impact on aggregate attendance over automatic promotion regime.

Model description		Impact on aggregate attendance	% increase over auto regime
Model 1	All variables	71569	0.91%
Model 2	No league positions	54595	0.69%
Model 3	nonzero prob only	153024	1.97%
Model 4	Promotion prob only	-19291	-0.24%
Model 5	Significance only	-14078	-0.18%

assumption that $\pi(prom|m)$ has no impact on $\pi_{it}(m)$.

$$\mu_{ijt}^{auto} = \exp(x_{it}^{auto}\beta_i + x_{jt}^{auto}\beta_j + z_t\beta_z + \bar{x}_{.t}^{po}\beta_{end} + v_i). \qquad (11.10)$$

Of particular note within this equation is the retention of the original values of $\bar{x}_{.t}^{po}$ rather than a new set $\bar{x}_{.t}^{auto}$. This is because the original values of $\bar{x}_{.t}^{po}$ (in combination with β_{end}) capture a component of a home team's time invariant heterogeneity rather than the impact of the play-off variables. The underlying heterogeneity is assumed to apply in different hypothetical settings and so should remain constant across designs.

In the 2000/01 season, the aggregate attendance across all teams was 7909514. Table 11.4 indicates the estimated incremental impact the play-off regime has made relative to what would have occurred had an automatic top-three promotion scheme been been in operation. In all five models, the incremental difference is rather modest ranging from an increase of 1.97% to a decrease of −0.24%.

Models 3 to 5 provide limits to the uncertainty attributable to the correlation between promotion related variables. Model 3 assumes that there is no impact of a redistribution of probability or significance and the only impact occurs through the creation of new nonzero probability games; this therefore presents the play-off system in its "best" light assuming that there are no counteracting effects of redistributing probability from large attendance generating teams to smaller teams. In this model, no team is worse off than under the automatic regime, and the overall impact is an additional 153024 attendees over the full regular season of 552 games, with an average of an additional 277 attendees per game. The increase is not even across teams. Teams such as Fulham (+1940), and Bolton (+2206) have increased aggregate home attendances through the impact on the away teams, whereas teams such as Norwich (+13689) and Sheffield United (+10783) gain more substantially by having far more nonzero probability games (both clubs had slim chances of obtaining a play-off place) and being reasonably big attendance producing clubs.

Model 4 represents the play-offs in their worst possible light, whereby the impact of the play-offs occurs through the redistribution of promotion probabilities. In this case a reduction of −0.24% attendance is estimated or a reduction of 35 fans per game. Again the reduction is not felt evenly across teams with Bolton (−28797) and Blackburn (−21343) losing the most over the season, reflecting the reduction in the probability of promotion for these two teams contesting second and third positions. Attendance during the post-season play-off games themselves may compensate for this loss; but for Blackburn, the team which eventually finished second, there would be no such compensating attendance. Fulham (−463) who only had relatively small probabilities of finishing third is thus only marginally affected. The teams that gained the most if this were the correct model are West Bromwich Albion (+10615) and Preston North End (+10626) — teams which spent much of the season in the lower play-off positions fifth and sixth.

The outcome of model 5 where the only consequence of play-off is assumed to be a redistribution of significance, is not as theoretically predictable as in models 3 and 4. The empirical estimate is actually a reduction of −0.18% or 26 attendees per game. The ambiguity of the anticipated effect is illustrated by the expected impact on Fulham (+2081). With a play-off system there is a significant difference between finishing second and third and little between third and fourth, whereas with an automatic promotion scheme there is a significant difference between finishing third and fourth and not between second and third. For Fulham, as they had a very low probability of finishing fourth, the shift of significance "up" the table meant that more of Fulham's games had greater overall significance. The opposite is true for Birmingham (−20981) which had a good chance of finishing third (they spent most of the season in fourth place). As the play-off system had removed much of the significance between finishing third and fourth, it thus removed much of the significance in Birmingham matches. Had the positions been reversed and larger club Birmingham had the season that smaller club Fulham had (and vice versa), the greater efficiency of Birmingham in turning significance into attendance may have reversed the overall picture. Bolton (−10168) is the other big loser whereas Sheffield United (+5311) and Preston (+4143) are the biggest gainers.

Models 1 and 2 include all play-off related variables and so allow for a net effect of the counteracting play-off related variables. They estimate the net impact of the play-off regime to be an increase in attendance of between 0.91% and 0.69%, an increase of between 71569 to 54595 or 130 to 99 attendees per game. This indicates the production of more nonzero probability games (keeping teams theoretically in contention) outweighs the redistribution of probability and significance from the larger to smaller clubs, at least in the season 2000/01. However, the overall effect is rather modest. Across teams, both models predict that Bolton, Birmingham Blackburn, Crewe, and Fulham (in model 2 only) lose attendance (Crewe via the impact on away teams), with Bolton (−27846), Blackburn (−15635), and Birmingham (−13793) being the major losers (though two of the three will gain attendance from the play-off games themselves), whereas all other teams gain smaller attendances, but on aggregate, sufficient to overcome the loses suffered by the bigger clubs.

11.8 CONCLUSIONS

The original research question aimed to analyze the extent to which post-season play-off systems influenced attendance at regular league matches. The analysis was applied to a single season of English league football and conducted in three steps:

construction of a simple theoretical model identifying play-off relevant parameters (promotion probability; nonzero probability; significance of match); a statistical estimate of these parameter values and the effect on attendance; and finally, a predication of attendances that would be observed given the values of the promotion variables implied by an automatic promotion regime. The theoretical model identifies that play-off related variables may act against each other and thus makes the impact of play-offs an empirical matter.

Section 11.3 identified the practical means of measuring the play-off related parameters in this analysis. Lack of ideal data has required construction of imperfect measures using ex-ante betting data in constructing the expectations of final league positions. This will probably introduce an element of right-hand side variable measurement error that is largest during the early part of the season (where the time difference between actual and ideal variable measurement is largest and expectations are a greater function of the simulation). While it is expected further research may target this issue and produce more refined measurements, inspection of matches which the flawed measure identify as significant feel intrinsically correct and incorporate current league positions, league structure, remaining fixtures, and some reasonable rational means of incorporating how the remaining fixtures will be resolved and how they will impact on the final table.

The empirical content of the paper is provided by analysis of the 2000/2001 English 1st Division season which is described in section 11.4. The use of a single season means that inference may be limited to that season alone and, in addition, may have contributed to issues of multicollinearity. However, the season does not look unduly exceptional, with larger teams finishing towards the top of the division, and so the results may be expected to replicated over other seasons. This descriptive section identifies the data issues involved — strictly nonzero, skewed and heteroscedastic attendance data clustered within teams as a result of unobserved heterogeneity between teams. The implications of which are discussed in section 11.5 and conclude that a GLM framework is appropriate to allow for the production of unbiased predictions of the impact on attendance, something the more common log transformation solutions do not permit. The GLM framework is supplemented with random effects and a correction for endogeneity. Within the model, a log link function allows clubs with different market sizes to have heterogeneous responses to determinants of demand and the choice of a Gamma distribution allows for unbiased estimates and predictions despite the skewed, nonzero, and heteroscedastic nature of the dependent variable. Though there does appear to be an issue with multicollinearity and the data may be limited in the extent it can separate out the individual effects of the elements of promotion probability and significance, sensitivity analysis shows the substantive overall results are generally robust on this issue. The estimation finds that two elements of match significance are consistently statistically significant: the nonzero probability of a game and the probability of promotion at the time of the game. The difference a match may make to the probability is of the correct sign but not significant unless the highly correlated variables are omitted from the specification.

The data also permits a detailed description of how the play-off system has reallocated probability and significance from bigger to smaller clubs leading to a potential for play-offs to actually reduce aggregate attendance. However, given the estimated relationship between game characteristics and attendance, the impact of creating more games where teams are theoretically still in contention for an end of season outcome is sufficient for play-offs to have increased overall attendance. In total the overall estimated impact is a rather modest increase of between 0.91% and 0.69% attendees during the regular season. This supports the notion that the English system has created additional attendance, but by an amount that may be smaller than anticipated. Two reasons stand out for this, firstly although statistically significant, the play-off related variables do not have a great practical impact on the attendance "production" function — a model which assumes no impact of the redistribution of probability still estimates a limited positive impact of just 1.79%. Such results may indicate the limitations other attendance generating policies may have. Secondly, although the net impact is positive, the redistribution effect reduces the overall gain, with the bigger teams occupying the higher play-off qualification positions (third and fourth) and second place losing amounts of attendance which are not matched by any other individual team gain and eclipsed only by the combined gain of all other teams.

REFERENCES

Berri, D.J., M.B. Schmidt, and S.L. Brook (2004). Stars at the gate: The impact of star power on NBA gate revenues. *Journal of Sports Economics 5*(1), 33–50.

Borland, J. and R. Macdonald (2003). Demand for sport. *Oxford Review of Economic Policy 19*, 478–502.

Cairns, J. (1990). The demand for professional team sports. *British Review of Economic Issues 12*(28), 1–20.

Dart, J. and J. Gross (2006, April 19). Does it really matter where you finish in the play-offs? *The Guardian.*

Duan, N. (1983). Smearing estimate: A nonparametric retransformation method. *Journal of the American Statistical Association 78*, 605–610.

Forrest, D. and R. Simmons (2002). Outcome uncertainty and attendance demand in sport: The case of English soccer. *The Statistician 51*(1), 13–38.

Garcia, J. and P. Rodriguez (2002). The determinants of football match attendance revisited: Empirical evidence from the Spanish football league. *Journal of Sports Economics 3*(1), 18–38.

Gardner, P. (2005). DC United outdo Adu. *World Soccer*, 22–23.

Greene, W.H. (2003). *Econometric Analysis* (5th ed.). Prentice Hall.

Kennedy, P. (2004). *A Guide to Econometrics* (5th ed.). Oxford (UK): Blackwell Publishing.

Kuypers, T. (1997). *The beautiful game? An econometric study of audiences, gambling, and efficiency in English football.* Ph. D. thesis, University College London, London.

Mabel's Tables (2003). Football yearbook. `http://www.mabels-tables.com`.

Maddala, G.S. (2001). *Introduction to Econometrics* (3rd ed.). New York: John Wiley.

Manning, W.G. and J. Mullahy (2001). Estimating log models: To transform or not to transform? *Journal of Health Economics 20*, 461–494.

McCullagh, P. and J.A. Nelder (1989). *Generalized Linear Models* (2nd ed.). Monographs on statistics and applied probability 37. London: Chapman and Hall.

Noll, R.G. (2003). The organization of sports leagues. *Oxford Review of Economic Policy 19*(4), 530–551.

Peel, D.A. and D.A. Thomas (1996). Attendance demand: An investigation of repeat fixtures. *Applied Economics Letters 3*(6), 391–394.

Rabe-Hesketh, S., A. Skrondal, and A. Pickles (2004). *GLLAMM Manual.* U.C. Berkeley Division of Biostatistics Working Paper Series.

Skrondal, A. and S. Rabe-Hesketh (2004). *Generalized Latent Variable Modeling: Multilevel, Longitudinal, and Structural Equation Models.* Boca Raton: Chapman & Hall/CRC.

StataCorp (2003). *Stata Statistical Software: Release 8.* College Station, TX: StataCorp LP.

Statmail (2003). Statmail. `http://www.statmail.co.uk`.

12

Measurement and interpretation of home advantage

Ray Stefani

California State University

ABSTRACT

Three factors appear to impart an advantage to the home team: a physiological factor due to travel fatigue that affects the visiting team, a psychological factor due to encouragement of the home team and intimidation of the visiting teams by the home crowd and a tactical factor due to familiarity of the home team with the venue. Three sports (rugby union at 25.1%, soccer at 21.7%, and NBA basketball at 21.0%) had the highest regular season home advantage measured by the fraction of home wins minus the fraction of home losses. The continuous nature of these sports is consistent with greater player fatigue than the other sports and logically greater home advantage for the more rested home team. Home advantage increased in soccer cup competition and NBA playoff competition by about 5% compared to the regular season. Three sports were at the middle of the range of home advantages: Australian Rules football at 18.8%, NFL football at 17.5%, and US college football at 16.6%. These three sports provided much more stoppage time and player substitution with resulting reduced excitation of the three factors. The playoff competition systems used in those three sports do not provide useful home advantage data. The lowest regular season home advantages were for NHL hockey at 9.7% and for MLB baseball at 7.5%. These sports also had the least excitation of the physiological, psychological, and tactical factors. NHL hockey playoff competition resulted in a lower home advantage than during the regular season. For MLB, playoff home advantage was about the same as for the regular season.

12.1   INTRODUCTION

One of the great pleasures of sport is to attend an event accompanied by family and friends and to cheer enthusiastically for a favorite local team. A casual review of the standings of most sports reveals the fact that a home team wins more than it loses and scores more points than the visiting opposition. What are the causes of the relative success of home teams? Do home teams have more advantage in some sports than others? How does playoff and cup home advantage compare to the regular season home advantage? These are some of the questions that I hope to answer in this section. A review of the work of Pollard (1986, 2002, 2006) and Stefani and Clarke

(1992) and others suggests three primary factors contributing to home advantage.

1. Travel fatigue of the visiting team (physiological factor);

2. Encouragement of the home team and intimidation of the visiting teams by the home crowd (psychological factor);

3. Familiarity of the home team with the venue (tactical factor).

The first task before us is to decide how to measure home advantage. Next, that measure of home advantage will be tabulated for eight sports and the relative home advantages of those sports will be compared to the relative influence of the three factors above. Regular season and playoff/cup post season home advantages will be compared. I conclude with a discussion of fairness and a summary.

12.2   MEASURING HOME ADVANTAGE

The term "home advantage" implies that a team performs better at home against an opponent than playing away against that same opponent. To quantify home advantage, two measures are commonly used: the difference in scoring and the difference in the fraction of games won. For our purposes, home advantage will be found in terms of the difference in the fraction of games won. That value is readily available using sports standings; hence, it is not necessary to search for home versus away scoring. We must take the relative strengths of the two teams into consideration. If a weak team plays at home and then away against a superior team, then the weak team may lose at both locations, which would seem to indicate that there is no home advantage, if we evaluate only a small number of such mismatched games.

A mathematical model is needed. For our purposes, each game is given a numerical index, ascending in the order played, and each team is also given a unique numerical index. Let us assume that every game index is called k so that the home team index is formally denoted $i(k)$, and the away team index is $j(k)$. To simplify notation, the functional dependence (k) is omitted. All summations are assumed to be taken over appropriate values of k for which there is a home team. Neutral field games are thus ignored. Let z_i denote a value $+1$ when the home team, designated i, wins in game k, zero for a home draw, and -1 for a home loss. Suppose a probit rating r varying from zero to one is found for each team, where each rating refers to team strength at a neutral venue. For convenience, additional variables are defined: w_i is one only for a home win, d_i is one only for a home draw, and l_i is one only or a home loss. The model used here is

$$z_i = r_i - r_j + h + e = w_i + 0d_i - l_i \qquad (12.1)$$

Note that the value of $r_i - r_j$ varies between $+1$ and -1, where a value of $+1$ predicts a neutral venue win for team i in the next game, a value of 0 predicts a neutral venue draw for team i, and a value of -1 predicts a neutral venue loss for team i. Here h quantifies the home advantage, and e is an error. Suppose that h is found to minimize

$\sum_k e^2$. Let the total number of home games be denoted

$$N = \sum_k (w_i + d_i + l_i) \tag{12.2}$$

The home advantage that minimizes $\sum_k e^2$ by smoothing past data is given by

$$h = \frac{1}{N} \sum_k (w_i + 0d_i - l_i) - \frac{1}{N} \sum_k (r_i - r_j) \tag{12.3}$$

The calculation of h is obviously coupled with finding the ratings, although we shall see that the calculation of h can be decoupled. Most of the competitions that follow involve a double round-robin schedule, in which all teams pair once at home and once away. For example, if team $i = 8$, having a rating of $r_i = 0.6$, plays at home against team $j = 5$, having a rating of $r_j = 0.4$, then $r_i - r_j = 0.2$. When the two teams switch locations, the home team becomes team $i = 5$ and the away team becomes team $j = 8$ so that $r_i - r_j = -0.2$. For the sum of all such paired games, $\sum_k (r_i - r_j)$ is zero. Of course, the ratings of two teams may change between competitions, but over an entire season, $\sum_k (r_i - r_j)$ should be small. For unbalanced competitions without uniform home-away pairs, if several seasons of data are taken, the home and away teams on average should have about equal ratings, so again, $\sum_k (r_i - r_j)$ would be approximately zero. Since we will be taking many seasons of data, it is reasonable to simplify Equation 12.3 to

$$h = \frac{1}{N} \sum_k (w_i - l_i) \tag{12.4}$$

Once h is calculated from Equation 12.4, the ratings may be found from any of a number of methods as in Stefani and Clarke (1992), Stefani (1987, 1998, 1999), and Clarke (1993). Note that a draw affects h only via adding into Equation 12.2 as a match played. If $N = 100$ matches are played, resulting in a total of 55 home wins, 10 draws and 35 home losses, then using Equation 12.4, h would be $0.01(55 - 35) = 0.20$ or 20%.

Table 12.1 compares home advantage h for eight sports: rugby union, soccer (called football in many nations), the National Basketball Association (NBA), Australian Rules football, the National Football League (NFL), USA college football, the National Hockey League (NHL), and Major League Baseball (MLB). Regular season data includes 98 seasons and 71,726 games. For some sports, playoff competition is included. For soccer, European international club cup competition is included. Altogether there are 46 seasons and 5,453 playoff and soccer cup matches included. See the legend in Table 12.1 for the range of seasons for each sport. Based on the home advantage values in Table 12.1, the eight sports fall into three groups.

All games listed in Table 12.1 must, of course, be played at the home team's venue. In Pollard (2002), the familiarity aspect of home advantage was evaluated via a study of 37 NBA, NHL, and MLB teams that changed stadiums. Home advantage was calculated in terms of points earned for the league table or standings, which is in

TABLE 12.1
Home Advantage Expressed as Home Wins Minus Home Losses
Rugby Union: Zurich Premiership (2002-03 to 2004-05), Super 12 (2001-05).
Soccer: Top Divisions in England and Italy (1994-95 to 2004-05); Germany and Norway (1994-95 to 1996-97); European Cup Competition (1994-95 to 2004-05). NBA (1992-93 to 2004-05). Australian Rules football (1980-1994). NFL (1997-98 to 2005-06). USA College football (1978-79 to 1980-81). NHL (1993-94 to 2003-04). MLB (1992-2005).

Sport	Regular Season		Playoff/Cup		Regular Season Home Advantage	Playoff/Cup Home Advantage (Adjusted Home Advantage)
	Seasons	Games	Seasons	Games		
Rugby Union	8	725			25.1%	
Soccer	28	9908	11	3304	21.7%	26.9%
NBA	12	13682	12	892	21.0%	29.6%(26.7%)
Australian Rules football	15	1681			18.8%	
NFL	9	2248			17.5%	
College football	3	1669			16.6%	
NHL	10	11466	10	861	9.7%	8.2%(7.2%)
MLB	13	30347	13	396	7.5%	7.1%
	98	71726	46	5453		

TABLE 12.2
Players, Ball in Play, and Substitutions for the Sports in Table 12.1.
NP = number of players per team.
PT = playing time in minutes for a regulation game.
GL = average game length in minutes to complete a regulation game.
$\%IP$ = percent of time the ball is in play = $100 \times PT/GL$.
NS = maximum number of substitutes available per game.
$SR?$ = Can a player that has been substituted for reenter the game?

Sport	NP	PT	GL	$\%IP$	NS	$SR?$
Rugby Union	15	80	105	76.2%	7	No
Soccer	11	90	109	82.6%	3	No
NBA Basketball	5	48	130	36.9%	Entire Team	Yes
Australian Rules Football	18	80	120	66.7%	4	Yes
NFL Football	11	60	190	31.6%	Entire Team	Yes
College Football	11	60	200	30.0%	Entire Team	Yes
NHL Hockey	6	60	140	42.8%	Entire Team	Yes
MLB Baseball	9		165	14.5%	Entire Team	No

In MLB Baseball, no official time is kept. There are approximately 24 minutes of significant action.

proportion to home advantage in terms of games won. The average home advantage became less in each sport in the first year of play in a new stadium. Pollard estimated that familiarity with the home venue accounts for about 25% of home advantage.

In order to compare the eight sports, Table 12.2 indicates the number of players per team, the playing time in minutes for a regulation game, the approximate game length in minutes required to complete those playing minutes including stoppage time and time between halves, quarters, and periods. Included also are the percent of time that the ball is in play (100×playing time/game length), the maximum number of substitutes allowed, and whether or not a player that has been substituted for can reenter the game.

12.3 RUGBY UNION, SOCCER, NBA

The highest regular season home advantages for the eight sports in Table 12.1 are for rugby union, soccer, and the NBA. Rugby union competition includes the Zurich Premiership and Super 12 competitions. The English professional rugby union championship was sponsored through the 2004-05 season by the Zurich Insurance Company; hence the name "Zurich Premiership." The 12 teams played a double round robin schedule. Starting with the 2005-06 season, the competition became the Guinness Premiership due to change of sponsorship. Through the 2004-05 season, Super 12 competition included 12 teams from Australia, New Zealand, and South Africa. The teams played a single round-round schedule so that home-away competition was unbalanced. Starting with the 2005-06 season, two teams were added, creating what is now Super 14 competition.

Soccer data covers the top divisions for England, Italy, Germany, and Norway. During the period of study there were 20-22 teams in the English Premier League, 18-20 teams in Italian Serie A, and 18 teams in the German First Bundesliga. Norwegian competition included the 12-14 team top division and the two 12 team leagues of the next division. All teams in those four countries played a double-round robin schedule. For the period of study, the NBA had 27-30 teams. Each opponent was not played against an equal number of times, but the 82-game schedule was balanced in that each opponent was played against an equal number of times home and away.

For the top three sports in Table 12.1, h for the regular season ranged from 21.0% to 25.1%. These sports have in common that action is primarily continuous, accentuating the physical disadvantage of travel fatigue. There is more physical contact in rugby union compared to the other two sports, consistent with greater fatigue and greater advantage for the home team. For rugby union competition with 15 players per team, the 80 playing minutes require about 105 clock minutes to complete a game, including about ten minutes of stoppage time and 15 minutes at half time. The ball is in play 76.2% of the time. Seven permanent substitutes are allowed per game. For soccer, with 11 players per team, three permanent substitutions are allowed per game. The 90 minutes of playing time require about 109 clock minutes, including about four minutes of stoppage time and 15 minutes at half time. The ball is in play 82.6% of the time. These are obviously fatiguing sports. In the NBA, with only five players per team to share the activity, compared to 11 and 15 in the other sports, fatigue is offset by the ability to substitute at any stoppage. Typically, the 48 minutes of playing time take about 130 clock minutes to complete, including various stoppages, so that the ball is in play 36.9% of the time.

A high fraction of time for the ball being in play tends to accentuate the psychological effects of crowd noise as play builds towards a score. As to tactical advantage, the home teams have better familiarity with the pitch surface in rugby and soccer, while NBA home players benefit from familiarity with background colors and shapes which would aid scoring. The short time to react due to the continuous nature of the three sports makes any such tactical advantage more significant than for a sport with many restarts following stoppages. In summary, the factors leading to home advantage are all enhanced by the continuous nature of the three sports, consistent with the highest home advantage during the regular season.

Another aspect of the influence of crowd passion upon home advantage in a continuous action sport, soccer, was examined by Nevill, Balmer, and Williams (2002). Referees were asked to view past game videos and to assign fouls. Those that viewed the videos with crowd noise assigned fewer fouls to home-team players than those that viewed the matches without crowd noise.

Home advantage may be differentially affected by the passion of some spectators. Table 12.3 disaggregates regular season home advantage for the four European soccer nations used in this study. The home advantages for Norway, England, and Germany vary only from 19.8% to 21.8% while home advantage for Italy is significantly higher at 24.8%. I believe the higher figure for Italy is an artifact of Italian history. Italy was not unified until 1870, prior to which the current land area was di-

TABLE 12.3
Regular Season Home Advantage in Soccer for the
Teams in Table 12.1.

Country		Regular Season	
	Seasons	Games	Home Advantage
Italy	11	3440	24.8%
Germany	3	918	21.8%
England	11	4262	19.8%
Norway	3	1288	19.8%
	28	9908	21.7%

TABLE 12.4
Super 12 Home Advantage for Domestic and
International Matches Five seasons:
2001-2005.

Type of Match	Games	Home Advantage
Domestic	93	12.9%
International	237	30.0%
	330	25.2%

vided into many small states. Italians remain loyal to their region, perhaps more than to the central government. Many of the soccer teams of the Italian top division, Serie A, were the principal cities of the former Italian states. To the usual local pride in a team, we can add loyalty to Italian ancestral heritage; hence there is a higher amount of crowd intimidation for a visiting team competing in a hostile environment.

The territoriality influence on home advantage in soccer was examined in Pollard (2006) with a similar conclusion. Home advantage in terms of league table points was compared for 72 countries. Pollard concluded that in Europe, home advantage was highest in the Balkan countries and lowest in Northern Europe. In South America, home advantage was highest in the Andean nations. In both Europe and South America, those nations with highest home advantage were those with a history of territorial conflict.

The influence of international travel is demonstrated in Table 12.4, which indicates home advantage for Super 12 domestic matches played in Australia, New Zealand, and South Africa compared to the international matches played by the same teams.

The home advantage is 30.0% for international matches compared to 12.9% for domestic matches. Coincident with the much higher home advantage for international matches is the much higher fatigue for time in transit as well as for time zones traversed (with inherent biological disruption) for international travel.

How does the home advantage for post-season competition compare to that for the regular season? For each of the two rugby union competitions, the regular season is followed by a four-team, single elimination playoff. For each of the three pairings, the higher ranked team plays one home game; hence true home advantage is mixed with bias-by-selection due to the better team playing at home. Playoff data are not included for rugby.

The best European clubs from a given season advance to international cup competition that takes the entire following season to complete. From 1993-94 through 1998-99 there were three such cup competitions: the Champions League, UEFA Cup, and Cup Winner's Cup. The Cup Winner's Cup was discontinued and merged into the UEFA Cup starting with the 1999-2000 season. Each round consisted either of a two-game home-away series or of a league-type competition with balanced home-away pairs. In Table 12.1, for the 3304 cup matches in soccer, home advantage was 26.9% compared to 21.7% for regular-season domestic competition in the four nations mentioned. The cup matches were played at midweek between weekend domestic league matches. Travel fatigue, crowd encouragement/intimidation and tactical advantage logically increase with travel into a foreign country. Correspondingly, home advantage increases.

For the NBA playoffs, 16 teams enter into a tournament of 15 playoff pairings (eight in the first round, followed by four, two, and then one) after which one champion remains. For the 1992-93 season through the 2001-02 season, the first round of eight pairings were contested on a best-of-five basis and the remaining seven series were on a best-of-seven basis. Starting with 2002-03, all 15 series became best-of-seven. Except for the championship series, the team with the better regular season record plays the first game at home in each pairing. In the championship series each year, the first home game alternates between the Eastern and Western finalist. It is possible for the home-away pattern to become unbalanced, in which case the last term on the right of Equation 12.3 might not be zero. It is then necessary to subtract $1/N \sum_k (r_i - r_j)$ from the observed value of Equation 12.4.

In a best-of-five series, the first team to win three games advances. The home team venue alternates on a 2-2-1 pattern (after two home games, the venue switches for two games and then returns to the first site). If the series ends after four games, each team plays a balanced home-away schedule. If the series ends after three or five games, the first home team has one extra home game. In that latter case, assuming equal numbers of three, four, and five game series, the effective value of $1/N$ in Equation 12.3 becomes $1/3 \times (1/3 + 1/5)$ or 0.178. How much benefit does the seeded first home team gain from an extra of home game? Based on the 2001-02 through 2004-05 playoffs, for rounds other than the championship round, the team that played at home first played 303 games and won 25% more games than it lost, thus $(r_i - r_j)$ was 25%. For a best of five series, it is necessary to adjust observed home advantage by $-0.178 \times 25\%$ or -4.45%.

There are two patterns for a best-of-seven series, in which the first team to win four games advances. In one pattern, the home team venue alternates on a 2-2-1-1-1 pattern (after two home games, the venue switches for two games and then the venue alternates after each game). The 2-2-1-1-1 pattern is used in NBA best-of-seven series other than the championship series. The other pattern follows a 2-3-2 order, which is used by the NBA in the championship round. If either patterned series ends after four or six games, both teams are at home an equal number of times. If the 2-3-2 series ends in five games, the first visitor gets an extra home game while if seven games are required, the first home team gets an extra home game. A 2-3-2 pattern balances home-away assignment. In that case, Equation 12.4 needs no adjustment. For the 2-2-1-1-1 pattern, if five or seven games are required, the first home team plays an extra home game. Given that about equal numbers of four, five, six, and seven game series are played, the effective value of $1/N$ becomes $1/4 \times (1/5 + 1/7)$ or 0.086. Notice that the effective $1/N$ is about twice as much for a best-of-five series as for a best-of-seven series; thus, the best-of-five pattern affords more benefit to the team playing at home in the first game. How much benefit does the seeded first home team gain from an extra home game? Using the value of $(r_i - r_j) = 25\%$, we have an adjustment of $-0.086 \times 25\%$ or -2.15%. Using the distribution of NBA playoff series over the period in question, it is necessary to adjust Equation 12.4 by -2.9%. The home advantage for the NBA playoffs was 29.6% compared to 21.0% for the regular season. If we subtract 2.9% from the 29.6% value, we get 26.7%, almost the same value as for soccer. The playoff home advantage for soccer and the NBA compare similarly to the regular season, in that the playoff values are each about 5% higher. The enhanced importance of the playoff games, the fatigue of the playoff schedule along with the size and closeness of the playoff crowds appears logically to increase home advantage due to the three factors noted earlier.

12.4 AUSTRALIAN RULES FOOTBALL, NFL, AND COLLEGE FOOTBALL

The middle group of sports in Table 12.1 include Australian Rules football (played by the Australian Football League, AFL), NFL football, and US College Football. During the seasons used, the 12-15 AFL teams played a 22 game schedule which created double round-robin play when there were 12 teams but unbalanced home-away pairs with expansion. Some teams played on shared grounds; hence, 26% of the matches were designated as having no home advantage when two teams sharing a home ground played each other. The 120 college teams played an unbalanced schedule of about 11 games each. The 30-32 NFL teams played a 16 game unbalanced schedule.

Home advantage in Table 12.1 varied from 16.6% for the NFL to 18.8% for the AFL, with college football being in the middle. How do these three sports differ from rugby union, soccer, and the NBA as to travel fatigue, crowd encouragement/intimidation, and tactical advantage, given that the middle three sports had less home advantage? Although Australian Rules football employs more continuous action than the other two football sports as the 18 players complete 80 minutes of playing time in about 120 clock minutes (with the ball being in play 66.7% of the time) and although substitution is limited to four players who can rotate in and out of action, the players do not follow the ball. Instead, the players are distributed across the length and breadth of the oval playing area. The ball is moved via kicking, punching, or running with the ball (which is dribbled every 15 meters somewhat like in basketball) while players remain in various sectors, reducing game fatigue and reducing the negative influence of travel fatigue. As to the other two football sports, every play is a restart where the 60 minutes of playing time require about 190 clock minutes for NFL competition (the ball being in play 31.6% of the time) and 200 clock minutes for college football where the ball is in play 30.0% of the time. Typically, there are three sets of players on each team that are substituted in mass throughout a game: the offensive, defensive, and special team players. The stoppage time and liberal substitutions reduce in-game fatigue and therefore reduce the detrimental effects of travel.

The large size of the playing oval in Australian Rules football probably reduces the crowd's psychological influence, compared to rugby union, soccer, and the NBA which also have high percentages for the ball being in play. The many play stoppages in the other two football sports cause crowd noise to build and then dissipate. These gaps in action can reduce the crowd's psychological influence. There is less variation among the playing surfaces in the middle three sports; hence, there is less tactical advantage when playing at home.

Playoff activity in each of Australian Rules football, NFL, and college football do not provide meaningful home advantage information. In Australian Rules football and in the NFL, a ladder playoff is used in which higher ranked teams play at home. Due to bias-by-selection, that playoff data is not useful. At the end of each regular college football season, "bowl games" are played. These are single games and the winners do not advance to further competition. Since most of these games are played at neutral venues, no useful home advantage information is conveyed.

12.5   NHL HOCKEY AND MLB BASEBALL

The lowest regular season home advantages in Table 12.1 are 9.7% for NHL hockey and 7.5% for MLB baseball. There were 26-30 NHL teams playing a balanced 82 game home-away schedule during the period of study and 26-30 MLB teams each playing a balanced 162 game schedule. Why would the NHL have a lower advantage than the NBA, when athletes of both sports have such a fatiguing duty cycle, given five and six players per team respectively? Certainly contact in the NHL is as violent as in rugby union, another sport with a high home advantage. A typical NHL game has 60 minutes of playing time, taking about 140 clock minutes to complete, given stoppage time, due in part to penalties and the puck leaving the rink, and time between periods. These delays terminate in about 60 face-offs per game. The puck is in play 42.8% of the time, compared to 36.9% for the NBA. An NHL team of six players will typically make substitutions using four "lines" of players about every 45 seconds of playing time. These substitutions significantly reduce in-game fatigue, thus reducing the differential effect of travel fatigue compared to the home team. The glass that surrounds the rink to protect spectators from a wayward puck also provides a psychological barrier from crowd encouragement/intimidation, if in fact a hockey player can be intimidated. As to tactical advantage for the home team, there are only minor differences in ice hardness and rebound tendency of the sideboards for a pass; therefore tactical advantage is minor for the home team.

Baseball with nine players per team has the lowest value of h, 7.5%, and is also by far the slowest paced sport of the eight. There is no fixed playing time. A regulation game consists of nine innings, in which the teams alternate batting with the visiting team batting first and the home team batting last for each of the nine innings. Each team is allowed three "outs" per inning, for a potential total of 27 outs each, per regulation game. An out in baseball is similar to a dismissal in cricket. Since the home team bats last, if the home team leads after the visiting team is finished in the ninth inning, the home team waives the last three outs. The average regulation game requires about 165 minutes for the 51 or 54 outs. During those 165 minutes, most activity involves just four players: the pitcher, catcher, batter, and perhaps one base runner. If we assume there are 250 pitches per game with 20 seconds per pitch and 2 minutes between 17 of the 18 team batting sessions, then there remain 48 minutes of action beginning with a pitch. If we assume significant action happens in about 24 of those 48 minutes, there is significant action in 24/165 of the time or 14.5%. Clearly, there is much less player fatigue, compared to the other sports. At important moments in the game, there can be significant crowd cheering. There are also some small tactical advantages for the home team due to familiarity with the home field layout, wind tendencies, and background colors that affect fielding and batting. The slow pace of the game allows adjustments to be made by the visiting team manager to reduce most of the home advantage. Further, travel expense is commonly reduced by scheduling three consecutive games at each venue visited. When the last two games are played in a given visit, travel fatigue has been reduced and the visitor has become familiar with the venue. The three factors leading to home advantage appear to be less activated in NHL and MLB competition, compared to the other five sports and home advantage is also less.

Although the home team normally bats last in baseball and wins more than it loses, there does not appear to be a tactical advantage due only to batting last. Courneya and Carron (1990) examined recreational softball in which there was neither an inherently larger crowd supporting the team batting last nor differential travel. They concluded that batting last did not convey an advantage. Bray, Obrara, and Kwan (2005) examined USA college NCAA baseball tournaments in which visiting teams played each other and again the conclusion was that batting last afforded no tactical advantage. If the team bats last and is behind, it is equally true that such a team can bring in "pinch hitters" to attempt to score the needed number of runs while the other team deploys a pitching specialist called a "closer" and aligns the fielders to reduce the likelihood of the known number of runs being scored. Simon and Simonoff (2006) similarly concluded no advantage for the team batting last in both NCAA men's baseball competition and NCAA women's softball competition. For close games, batting last in softball showed some disadvantage, possibly due to the dominance of softball pitching over softball hitting.

How does the home advantage in the playoffs compare to the regular season? Over the study period, the NHL playoff system employed 15 pairings for the 16 teams on a best-of-seven, 2-2-1-1-1 basis. As discussed for NBA competition, the effective value of $1/N$ is 0.086 for a 2-2-1-1-1 series. Based on NHL playoff series for the 2000-01 through 2003-04, the higher seeded teams played 330 games and won 13% more games than they lost. The championship series is not seeded, so the correction is $-(14/15) \times 0.086 \times 13\%$ or -1%. For the NHL h was 8.2% for the playoffs. After adjustment, h becomes 7.2%, 2.5% lower than for regular season play.

For the first three years of this study, MLB used four playoff teams who were paired in three best-of-seven series using the 2-3-2 pattern. For the more recent four years of this study, MLB used eight playoff teams. The first four pairings were of the best-of-five 2-2-1 pattern, while the last three series were of the 2-3-2 pattern. For the 2000 though 2005 playoffs, higher seeded teams won 85 games and lost 84. The rightmost term in Equation 12.4 is then near zero and no adjustment is needed. For MLB, playoff h was 7.1%, 0.4% less than the value for the regular season; thus there was no significant difference.

In the NHL, perhaps the three factors contributing to home advantage are further diminished due to the length of the playoff period in which more than one game may be played at one venue. For MLB, with the lowest regular season home advantage, playoff conditions are apparently the same as for the regular season which already employs extended competition at each venue.

12.6   CAN HOME ADVANTAGE BECOME UNFAIR?

For an excellent coverage of fair play in sport, see Loland (2001). I suggest a simple test for fairness in the context of home advantage. At the start of this chapter, I mentioned the pleasure of attending a game and cheering enthusiastically for a favorite local team. When that attendance is no longer a pleasure due to hooliganism, drunkenness, the shouting of racial epithets, and the hurling of objects toward the field, arguably those activities do not promote a fair home advantage. If you would consider it hypocritical to tell a young spectator that cheating on a test is not acceptable while your home team has altered the playing surface to gain advantage over the visiting team, then I suggest that such activity is not fair. Put more succinctly, a home team should provide the very same environment for a visitor that they themselves would expect as a visitor. Spectators for both teams should be able to enjoy the competition in the spirit of fair play.

12.7   SUMMARY

Three factors appear to impart an advantage to the home team. These are: a physiological factor due to travel fatigue that affects the visiting team, a psychological factor due to encouragement of the home team and intimidation of the visiting teams by the home crowd, and a tactical factor due to familiarity of the home team with the venue. The relative amount of fatigue in a given sport is the easiest to evaluate. The influence of the other two factors is much more subtle. Three sports (rugby union at 25.1%, soccer at 21.7%, and NBA basketball at 21.0%) had the highest regular season home advantage measured by the fraction of home wins minus the fraction of home losses. The continuous nature of these sports is consistent with greater player fatigue than the other sports and logically greater home advantage for the more rested home team. Arguably, crowd encouragement/intimidation and tactical advantage are enhanced with continuity of action. The influence of a passionate home crowd was evidenced by domestic soccer competition in Italy which had a much higher home advantage at 24.8% than for Germany (21.8%), England (19.8%), and Norway (19.8%). Home advantage increased in soccer cup competition and NBA playoff competition by about 5% compared to the regular season, after adjusting NBA playoff home advantage by 2.9% to compensate for the effect of best-of-five and best-of-seven playoff series patterns. Arguably, the three factors leading to home advantage were more activated in such post season play.

Three sports were at the middle of the range of home advantages: Australian Rules football at 18.8%, NFL football at 17.5%, and US college football at 16.6%. These three sports provided much more stoppage time and player substitution with resulting reduced excitation of the three factors than for rugby union, soccer, and NBA basketball. The playoff competition systems used in those three sports do not provide useful home advantage data.

The lowest regular season home advantages were for NHL hockey at 9.7% and for MLB baseball at 7.5%. These sports also had the least excitation of the physiological, psychological, and tactical factors. NHL hockey playoff competition resulted in a lower home advantage than during the regular season by about 2.5%, after adjusting for the effect of the best-of-seven playoff pattern. For MLB, playoff home advantage was about the same as for the regular season.

REFERENCES

Bray, S.R., J. Obrara, and M. Kwan (2005). Batting last as a home advantage factor in men's NCAA tournament baseball. *Journal of Sports Sciences 23*(7), 681–686.

Clarke, S.R. (1993). Computer forecasting of Australian Rules Football for a daily newspaper. *Journal of the Operational Research Society 44*, 753–799.

Courneya, K.S. and A.V. Carron (1990). Batting first versus last: Implications for the home advantage. *Journal of Sport and Exercise Psychology 12*, 312–316.

Loland, R. (2001). *Fair Play in Sport*. London: Routledge.

Nevill, A.M., N.J. Balmer, and A.M. Williams (2002). The influence of crowd noise and experience upon refereeing decisions in football. *Psychology of Sport and Exercise 3*, 261–272.

Pollard, R. (1986). Home advantage in soccer: A retrospective analysis. *Journal of Sports Sciences 4*, 237–248.

Pollard, R. (2002). Evidence of a reduced home advantage when a team moves to a new stadium. *Journal of Sports Sciences 20*(12), 969–973.

Pollard, R. (2006). Worldwide regional variations in home advantage in association football. *Journal of Sports Sciences 24*(3), 231–240.

Simon, G.A. and J.S. Simonoff (2006). "Last licks": Do they really help? *The American Statistician 60*(1), 13–18.

Stefani, R.T. (1987). Applications of statistical methods to American Football. *Journal of Applied Statistics 14*(1), 61–73.

Stefani, R.T. (1998). Predicting outcomes. In J. Bennett (Ed.), *Statistics in Sport*, Chapter 12. New York: Oxford University Press.

Stefani, R.T. (1999). A taxonomy of sports rating systems. *IEEE Trans. On Systems, Man and Cybernetics, Part A 29*(1), 116–120.

Stefani, R.T and S.R. Clarke (1992). Predictions and home advantage for Australian rules football. *Journal of Applied Statistics 19*(2), 251–259.

13

Myths in Tennis

Jan Magnus

Tilburg University

Franc Klaassen

University of Amsterdam

ABSTRACT

Many people have ideas about tennis. In particular, most commentators hold strong ideas about, for example, the advantage of serving first in a set, the advantage of serving with new balls, and the special ability of top players to perform well at the "big" points. In this chapter we shall investigate the truth (more often the falsity) of a number of such hypotheses, based on Wimbledon singles' data over a period of four years, 1992–1995.

13.1 INTRODUCTION

The All England Croquet Club was founded in the Summer of 1868. Lawn tennis was first played at The Club in 1875, when one lawn was set aside for this purpose. In 1877 The Club was retitled The All England Croquet and Lawn Tennis Club and the first tennis championship was held in July of that year. Twenty-two players entered the event which consisted of men's singles only. Spencer Gore became the first champion and won the Silver Challenge Cup and twelve guineas, no small sum (about £800 in today's value), but rather less than the £655000 that the 2006 champion Roger Federer received. In 1884 the women's singles event was held for the first time. Thirteen players entered this competition and Maud Watson became the first women's champion receiving twenty guineas and a silver flower basket. (William Renshaw, the 1884 men's singles champion received thirty guineas.) In 1922 The Championships moved from Worple Road to its current location at Church Road; see Riddle (1988) and Little (1995) for some historical details. For more than a century The Championships at Wimbledon have been the most important event on the tennis calendar. Currently both the men's singles and the women's singles events are restricted to 128 players.

Because of television broadcasts, tennis has become a sport which is viewed by millions all over the world. Many have ideas about tennis. In particular, most commentators hold strong ideas about, for example, the advantage of serving first in a set, the advantage of serving with new balls, and the special ability of top players to perform well at the "big" points. In this chapter we shall investigate the truth (more often the falsity) of a number of such hypotheses.

In Section 13.2 we discuss the data and some selection issues. Our data are obtained from Wimbledon singles' matches over a period of four years, 1992–1995. Further data at this level of detail were not available to us. In Section 13.3 we discuss three popular myths concerning the service and show that all three are false. The first two myths were earlier discussed in Magnus and Klaassen (1999c), and the third in Magnus and Klaassen (1999a). Section 13.4 discusses the notion of "winning mood" (dependence, in statistical parlance) by considering the final set and breaks. The two myths about the final set were analyzed in Magnus and Klaassen (1999b); the two myths about breaks are new results. Section 13.5 concerns "big points" (that is, identical distribution for the statistician); these results are also new.

We shall see that almost all hypotheses are rejected, but not all. It is not true that serving with new balls or starting to serve in a new set are advantages. Also, the seventh game is not especially important. But, it is true that big points exist and that real champions perform their best at such points.

13.2   THE DATA AND TWO SELECTION PROBLEMS

We have data on 481 matches played in the men's singles and women's singles championships at Wimbledon from 1992 to 1995. This accounts for almost half of all singles matches played during these four years. For each of these matches we know the exact sequence of points. We also know at each point whether or not a second service was played, and whether the point was decided through an ace or a double fault. Table 13.1 provides a summary of the data. We have slightly more matches for men than for women, but of course many more sets, games, and points for the men's singles than for the women's singles, because the men play for three sets won and the women for two. The men play fewer points per game than the women, because the dominance of their service is greater. But the women play fewer games per set on average—scores like 6–0 and 6–1 are more common in the women's singles than in the men's singles—because the difference between the 16 seeded and the 112 non-seeded players is much greater. (Until 2000, 16 players out of 128 were seeded at Wimbledon, thereafter 32. Seeded players receive a protected placing on the schedule, so that they cannot meet in the early stages of the tournament. Typically, but not necessarily, the top players in the world rankings are seeded.) This also leads to fewer tie-breaks in nonfinal sets for women. (By "final set" we mean the fifth set in the men's singles and the third set in the women's singles. At Wimbledon there is no tie-break in the final set.) Both men and women play about 60 points per set. The men play on average 230 points per match, the women 132, and hence a match in the men's singles takes on average 1.75 times as long as a match in the women's singles.

All matches in our data set are played on one of the five "show courts": Centre Court and courts 1, 2, 13, and 14. Usually matches involving top players are scheduled on these courts. This causes an under-representation in the data set of matches involving nonseeded players. This is, however, not the only selection problem. If two nonseeded players play against each other in the quarter-final, this match is likely to

TABLE 13.1

Number of matches, sets, games, tie-breaks, and points in the data set.

	Men's singles	Women's singles
Matches	258	223
Nonfinal sets	899	446
Final sets	51	57
Games	9367	4486
Tie-breaks	177	37
Points	59466	29417
Averages		
Sets in match	3.7	2.3
Games in nonfinal set	9.8	8.9
Games in final set	11.1	9.2
Tie-breaks in nonfinal set	0.2	0.1
Points in match	230.5	131.9
Points in game	6.1	6.5
Points in tie-break	12.1	11.8

be scheduled on a show court. But, if they play in the first round, their match is considered to be of less importance and is likely to be played on another court. After all, there are 16 first-round matches involving a seeded player and such matches usually take precedence. Therefore, the under-representation of matches between two nonseeded players is most serious in early rounds. This dependence on round in the selection of matches is also present in other types of matches, although it is less serious, as Table 13.2 shows.

We distinguish between round (1, first round; 7, final) and type of match (Sd-Sd for two seeded players, Sd-NSd for a seeded against a nonseeded player and NSd-NSd for two nonseeded players). The column labeled "Sam" in each part contains the number of matches in our sample, and the column labeled "Pop" the number of matches actually played. (Note that in the first round of the women's singles there are 63 rather than 64 matches between a seeded and a nonseeded player. The reason is that Mary Pierce, seeded 13, withdrew in 1993 at the last moment. She was replaced by Louise Field, an unseeded player.)

We see that the percentage of matches of nonseeded against nonseeded (NSd-NSd) players in our data set is 24.9 (74/297) for the men and 14.8 (42/283) for the women. Both are lower than the percentages for Sd-NSd matches, which are themselves lower than those for Sd-Sd matches. This illustrates the first selection problem, namely the under-representation of matches involving nonseeded players.

The second selection problem, caused by the round dependence, results from the increasing pattern in the sampling percentages over the rounds. For example, only 32.0% (82/256) of all first-round matches in the men's singles and 26.2% (67/256) in the women's singles are in the data set, whereas all finals have been sampled.

TABLE 13.2
Number of matches in the sample (Sam) and in the population (Pop).

Round	Sd-Sd		Sd-NSd		NSd-NSd		Total	
	Sam	Pop	Sam	Pop	Sam	Pop	Sam	Pop
(a) Men's singles								
1	—	—	48	64	34	192	82	256
2	—	—	46	54	16	74	62	128
3	—	—	39	41	16	23	55	64
4	8	9	15	15	8	8	31	32
5	7	7	9	9	0	0	16	16
6	7	7	1	1	0	0	8	8
7	4	4	0	0	0	0	4	4
Total	26	27	158	184	74	297	258	508
(b) Women's singles								
1	—	—	43	63	24	193	67	256
2	—	—	43	58	3	70	46	128
3	—	—	42	48	12	16	54	64
4	8	8	20	21	2	3	30	32
5	11	12	3	3	1	1	15	16
6	6	6	1	2	0	0	7	8
7	4	4	0	0	0	0	4	4
Total	29	30	152	195	42	283	223	508

Since we wish to make statements about Wimbledon (and not just about the matches in our sample), we account for both selection problems by weighting the matches when computing the statistics in this chapter. The weights are calculated as the ratios Pop/Sam in Table 13.2. This procedure involves an assumption, namely that within each cell the decision by Wimbledon's organizers whether a match is on a show court or not is random, so that the matches on the show courts (which are the matches that we observe) are representative. One could argue that, if the sample is very small compared with the population, this method would make the few observed matches too important. Most notably, in the women's singles we observe only three of the 70 matches played between two nonseeded players in the second round. If these three matches were selected by the organizers to include, for example, players just outside the top 16, then our method would be seriously biased for this cell. As it happens, the three matches concern players with Women's Tennis Association rankings 27-41, 131-143, and 22-113 and hence there is no reason to believe that these matches are not representative.

13.3   SERVICE MYTHS

The service is one of the most important aspects of tennis, particularly on fast surfaces such as the grass courts at Wimbledon. In Table 13.3 we provide some of its characteristics. As before, Sd-NSd indicates a match of a seeded against a nonseeded player, where the first player (Sd) is serving in the current game and the second player (NSd) receiving. Sd-Sd, NSd-Sd, and NSd-NSd are similarly defined. Standard errors are given in parentheses. To obtain the standard errors, we have treated all points as independent. This is not quite true (see Klaassen and Magnus, 2001): a good point (for example, an ace) may bring about another good point, and a bad point (missed smash) may bring about another bad point; but it is sufficient as a first-order approximation for our purpose.

We see that, at Wimbledon, the men serve almost three times as many aces as the women, but serve the same number of double faults. (The percentage of aces is defined as the ratio of the number of aces (first or second service) to the number of points served, rather than to the number of services.) In understanding the other service characteristics in Table 13.3, the distinction between "points won if 1st (2nd) service in" and "points won on 1st (2nd) service" is important. In the men's singles, when two seeded players play against each other, the first service is in 58.7% of the time. If the first service is in, the probability of winning the point is 77.7%. Therefore, in the men's singles the probability of winning the point on the first service is 58.7% $\times$ 77.7% = 45.6%; see the second column of Table 13.3, part (a). Hence,

% points won on 1st service

$$= (\% \text{ points won if 1st service in}) \times (\% \text{ 1st services in}), \qquad (13.1)$$

and, of course, the same for the second service. Combining the data for the first and second services, we can derive the percentage of points won on service. A player can win a point on service in two ways: on the first or on the second service. However,

TABLE 13.3
Service characteristics.

Characteristic	% of the characteristics for:				
	Sd-Sd	Sd-NSd	NSd-Sd	NSd-NSd	Total
(a) Men's singles					
Aces	11.7	11.0	7.7	7.0	8.2
	(0.4)	(0.2)	(0.2)	(0.2)	(0.1)
Double faults	5.1	5.1	5.8	5.6	5.5
	(0.3)	(0.2)	(0.2)	(0.2)	(0.1)
Points won on service	67.0	69.3	61.1	63.7	64.4
	(0.6)	(0.4)	(0.4)	(0.4)	(0.2)
1st services in	58.7	59.6	59.4	59.5	59.4
	(0.6)	(0.4)	(0.4)	(0.4)	(0.2)
2nd services in	87.8	87.3	85.6	86.2	86.4
	(0.6)	(0.4)	(0.4)	(0.4)	(0.2)
Points won if 1st service in	77.7	78.1	70.2	72.4	73.3
	(0.7)	(0.4)	(0.4)	(0.5)	(0.2)
Points won if 2nd service in	59.0	64.5	55.6	59.2	59.4
	(1.0)	(0.6)	(0.6)	(0.6)	(0.3)
Points won on 1st service	45.6	46.5	41.7	43.1	43.6
	(0.6)	(0.4)	(0.4)	(0.4)	(0.2)
Points won on 2nd service	51.8	56.3	47.6	51.0	51.4
	(0.9)	(0.6)	(0.6)	(0.6)	(0.3)
Games won on service	86.0	88.9	74.1	79.7	80.8
	(1.1)	(0.6)	(0.8)	(0.8)	(0.4)
(b) Women's singles					
Aces	3.3	4.2	2.5	2.9	3.1
	(0.3)	(0.2)	(0.2)	(0.2)	(0.1)
Double faults	3.9	4.2	5.8	6.0	5.5
	(0.3)	(0.2)	(0.2)	(0.3)	(0.1)
Points won on service	56.9	62.9	50.1	55.8	56.1
	(0.8)	(0.5)	(0.5)	(0.7)	(0.3)
1st services in	65.5	61.5	60.5	60.2	60.8
	(0.7)	(0.5)	(0.5)	(0.7)	(0.3)
2nd services in	88.8	89.1	85.2	85.1	86.0
	(0.8)	(0.5)	(0.6)	(0.8)	(0.3)
Points won if 1st service in	62.5	69.6	56.4	61.7	62.2
	(0.9)	(0.6)	(0.6)	(0.8)	(0.4)
Points won if 2nd service in	51.8	58.6	47.4	55.2	54.1
	(1.4)	(0.9)	(0.9)	(1.1)	(0.5)
Points won on 1st service	41.0	42.8	34.1	37.1	37.8
	(0.8)	(0.5)	(0.5)	(0.6)	(0.3)
Points won on 2nd service	46.0	52.2	40.4	47.0	46.6
	(1.3)	(0.8)	(0.8)	(1.1)	(0.5)
Games won on service	66.5	77.8	49.4	62.8	63.4
	(1.9)	(1.1)	(1.3)	(1.6)	(0.7)

the second possibility only becomes relevant when the first serve is a fault. Thus,

% points won on service = % points won on 1st service

$$+ (\% \text{ 1st service not in}) \times (\% \text{ points won on 2nd service}). \quad (13.2)$$

For example, from the second column of Table 13.3, part (a),

$$67.0\% = 45.6\% + (100 - 58.7)\% \times 51.8\%.$$

The literature on the service in tennis concentrates on the first/second service strategy rather than on myths relating to the service. In Gillman (1985) it is suggested that "missing more serves may win more points;" see also Gale (1980), George (1973), Hannan (1976), and Norman (1985). Borghans (1995) analyzed the 1995 Wimbledon final between Sampras and Becker, and showed that Becker could have performed much better if he had put more power in his second service. In a recent paper, Klaassen and Magnus (2006) ask whether the service strategy is efficient and measure its inefficiency.

13.3.1 A PLAYER IS AS GOOD AS HIS OR HER SECOND SERVICE

Many commentators say this, but is it true? Let us first ask how to measure the quality of the second service. This is not the percentage of second services in or the percentage of points won if the second service is in, but it is the combination of the two, namely the percentage of points won on the second service. In the men's singles, let us compare the matches Sd-Sd and NSd-NSd; see the second and fifth columns in Table 13.3, part (a). With some simplification, the players in these matches can be considered to have the same strength: they are either both good (NSd-NSd) or both very good (Sd-Sd). We see that the seeded players win significantly more points on their first service than do the nonseeded players (45.6% > 43.1%), but that the estimated probabilities of winning points on the second service are not significantly different. (Throughout this chapter we use a 5% level of significance.)

Hence, a seeded player distinguishes himself from a nonseeded player by having a better first service, not by having a better second service. Therefore, the idea that "a player is as good as his or her second service" is not supported by the Wimbledon data; The same holds in the women's singles as can be verified from part (b) of Table 13.3.

There is, however, one important difference between the men's singles and the women's singles and this relates to the quality of the first service. Referring to Equation (13.1), the quality of the first service is made up of two components: the percentage of points won if the first service is in and the percentage of first services in. In the men's singles the difference in the quality of the first service between seeded and nonseeded players is determined primarily by the percentage of points won if the first service is in (77.7% is significantly larger than 72.4%), whereas the difference in the percentage of first services in is not significant. In the women's singles the difference is determined primarily by the percentage of first services in (65.6% is significantly larger than 60.2%), whereas the difference in the percentage of points won if first service is in is insignificant.

We conclude that it is not true that "a player is as good as his or her second service." It would be more realistic to say that "a player is as good as his or her *first* service." The first service is more important than the second, but the aspect of the first service which matters differs between men and women.

13.3.2 SERVING FIRST

Most players, when winning the toss, elect to serve. Is this a wise strategy? This depends on whether or not you believe that there exists a psychological advantage to serve first in a set. This idea is based, presumably, on the fact that the player who receives in the first game is usually one game behind and that this would create extra stress. The advantage, if is exists, can only be psychological, because there is no theoretical advantage (Kingston, 1976; Anderson, 1977). Let us investigate whether there is any truth in this *idée reçue*.

Our first calculations seem to indicate that the idea must be wrong. Overall only 48.2% of the sets played in the men's singles are won by the player who begins to serve in the set. In the women's singles the percentage is 50.1%. The standard errors of the two estimates are 1.6% and 2.2%, respectively. Therefore, neither of the two percentages is significantly larger than 50%. If we look at the sets separately, then we see that this finding (starting to serve provides no advantage) seems to be true in every set, except perhaps the first. In the men's singles the estimated probability of winning a set when starting to serve is 55.4% (3.1%) in the first set and 44.3% (3.1%), 43.5% (3.1%), 51.0% (4.5%), and 48.8% (7.0%) in the second to fifth sets respectively. In the second and third sets serving first may even be a disadvantage.

Exactly the same occurs in the women's singles. There the probability that the player who starts to serve also wins the set is estimated to be 56.6% (3.3%) in the first set, 44.0% (3.3%) in the second set, and 47.8% (6.6%) in the third set. Apparently starting to serve in a set is a *disadvantage* rather than an advantage, except perhaps in the first set.

This is a little puzzling. We can accept perhaps that there is no advantage in serving first, but why should there be a disadvantage, and why should this disadvantage only exist in second and following sets, but not in the first? Let us take a closer look.

Consider a match between a stronger player and a weaker player, and consider the beginning of the second set. It is likely that the stronger of the two players has won the first set; after all he/she is the stronger player. Also, it is likely that the last game of the previous set has been won by the server in that game. As a result, the stronger player is likely to have served out the set, and hence the weaker player will (usually) start serving in the second set. The probability that the player who starts to serve in the second set will more often lose than win the set may therefore be due to the fact that it is typically the weaker player who starts to serve in the second set. The same holds for all sets, except the first. This explains that in all sets except the first the percentages are less than 50%, not because there is a disadvantage for the player who serves first in a set, but because the server in the first game is usually the weaker player. A proper analysis should take this into account. and this calls for a conditional rather than an unconditional analysis.

TABLE 13.4
Estimated probabilities of winning a set after winning the previous set (S: starts serving; R: starts receiving).

Set	Sd-Sd		Sd-NSd		NSd-Sd		NSd-NSd		Total	
	S	R	S	R	S	R	S	R	S	R
(a) Men's singles										
1	46.8	53.2	80.6	78.9	21.1	19.4	56.9	43.1	55.4	44.6
	(9.8)	(9.8)	(4.2)	(4.9)	(4.9)	(4.2)	(5.8)	(5.8)	(3.1)	(3.1)
2	61.0	52.6	82.1	78.8	19.3	30.8	72.1	70.0	72.5	68.0
	(21.8)	(10.9)	(6.0)	(4.4)	(16.1)	(9.1)	(9.0)	(6.6)	(5.1)	(3.5)
3	79.5	75.2	73.5	76.7	64.2	40.5	75.0	73.6	73.9	72.1
	(12.8)	(10.8)	(6.7)	(4.7)	(13.8)	(10.5)	(9.2)	(6.1)	(4.7)	(3.4)
4	19.5	50.0	74.4	69.8	26.2	34.3	75.9	68.8	62.9	60.2
	(17.7)	(14.4)	(11.3)	(10.3)	(10.1)	(10.4)	(10.4)	(11.6)	(6.5)	(5.9)
5	0.0	11.9	70.5	86.1	24.8	37.1	52.8	60.6	48.3	51.0
	(0.0)	(11.5)	(26.3)	(13.1)	(21.6)	(13.4)	(17.7)	(18.5)	(12.5)	(8.5)
(b) Women's singles										
1	62.4	37.6	72.4	84.0	16.0	27.6	64.3	35.7	56.6	43.4
	(9.0)	(9.0)	(5.1)	(4.3)	(4.3)	(5.1)	(7.4)	(7.4)	(3.3)	(3.3)
2	64.3	72.6	89.5	90.3	27.2	31.9	69.5	74.3	72.0	75.2
	(14.5)	(10.5)	(4.3)	(3.6)	(12.3)	(10.2)	(10.3)	(9.3)	(4.6)	(3.8)
3	40.4	25.0	67.5	85.9	0.0	15.0	92.9	60.8	63.5	60.1
	(21.9)	(21.7)	(14.8)	(9.7)	(0.0)	(13.5)	(12.9)	(17.3)	(9.6)	(8.9)

In part (a) of Table 13.4 we consider a player in the men's singles *who has won the previous set* and compare the estimated probability that he wins the current set when starting to serve with the estimated probability that he wins the current set when starting to receive. For example, if a seeded (Sd) player has won the first set against a nonseeded (NSd) player, then his probability of winning the second set is estimated as 82.1% when he (the seeded player) begins to serve and as 78.8% when his opponent begins to serve. Of course, there is no set before the first and hence the probabilities in the first row are simply the (unconditional) probabilities of winning the first set. The same probabilities, estimated for the women's singles, are provided in part (b) of Table 13.4.

Let us consider the first three sets in the men's singles and the first two sets in the women's singles, because the other sets have relatively few observations and hence large standard errors. There is some indication that in the men's singles there is an advantage in serving first: the overall probability of winning a set after winning the previous set is higher for the player who begins to serve than for the player who begins to receive. The difference is 10.8 percentage points (6.2%) in the first set, 4.5 percentage points (6.2%) in the second set, and 1.8 percentage points (5.8%) in the third set. However, these results are not significant and the differences are not positive for all four subcategories (Sd-Sd, Sd-NSd, etc.). We conclude that the support for our hypothesis is insufficient, except perhaps in the first set.

In the women's singles the first set indeed appears to be special. The probability of winning the first set is significantly higher for the player who begins to serve than for the player who begins to receive (the difference is 13.2 percentage points on average with a standard error of 6.6%). The probability of winning the second set after winning the first is lower, although not significantly, for the player who begins to serve than for the player who begins to receive (the difference is 3.2 percentage points with a standard error of 6.0%). Hence, in the women's singles our hypothesis is only true for the first set. This implies that electing to serve after winning the toss is generally better than electing to receive.

There is only one result in this analysis that is reasonably robust and that is that in the first set of a match there is an advantage in serving first. But why should the first set be different from other sets? Maybe this is because fewer breaks occur in the first few games of a match? Yes, but we can be more precise. The reason why the probability in the first set is higher is entirely due to the effect of the first game in the match. The probability that the server wins this first game is 87.7% (2.0%) rather than 80.8% in the men's singles and 74.3% (2.9%) rather than 63.4% in the women's singles; see Table 13.3. It is only the very first game that is special. In the second game the percentages of winning a service game are not significantly different from those concerning the first set excluding the first two games. This holds for both men and women.

Hence, what to do when you win the toss? Is it wise—as sometimes argued—to elect to *receive* when you win the toss, because it is easier to break your opponent in the first game of the match than in later games. In the first game, the argument goes, the server is not yet playing his or her best tennis. If this is true, however, then it must be that the receiver is performing even worse in the first game, as we find that breaks in the first game occur less often. Therefore, it is wise to elect to *serve* when you win the toss.

13.3.3 NEW BALLS

Tennis as a game has a long history which goes back to the Greeks and Romans. But it was not until 1870 that it became technically possible to produce rubber balls which bounce well on grass. When the All England Lawn Tennis Club decided to hold their first championships in 1877, a three-man subcommittee drew up a set of laws. Rule II stated that

> "the balls shall be hollow, made of India-rubber, and covered with white cloth. They shall not be less than 2 1/4 inches, nor more than 2 5/8 inches in diameter; and not less than 1 1/4 ounces, nor more than 1 1/2 ounces in weight." (Little, 1995, p. 284)

The quality of the tennis balls has gradually improved. From 1881 to 1901 the balls were supplied by Ayres; thereafter by Slazinger and Sons. Yellow balls were introduced at the 100th championships meeting in 1986. During the 1877 championships 180 balls were used; now more than 30000 are used in one year.

During a tennis match new balls are provided after the first seven games (to allow for the preliminary warm-up) and then after each subsequent nine games. Most commentators and many spectators believe that new balls are an advantage to the server. But is this true?

TABLE 13.5
Service characteristics depending on the age of the balls.

Characteristic	Percentage of the characteristics for the following ages of balls:									Total
	1	2	3	4	5	6	7	8	9	
(a) Men's singles										
Aces	8.7	7.9	8.6	7.7	8.2	8.3	8.5	8.4	7.2	8.2
	(0.4)	(0.4)	(0.3)	(0.3)	(0.3)	(0.3)	(0.3)	(0.4)	(0.3)	(0.1)
Double faults	5.8	5.3	5.8	6.6	5.4	5.6	4.9	5.1	5.1	5.5
	(0.3)	(0.3)	(0.3)	(0.3)	(0.3)	(0.3)	(0.3)	(0.3)	(0.3)	(0.1)
Service points won	64.7	63.4	64.2	64.8	64.0	64.1	65.8	64.3	64.3	64.4
	(0.6)	(0.6)	(0.6)	(0.6)	(0.6)	(0.6)	(0.6)	(0.6)	(0.6)	(0.2)
1st service in	58.9	60.2	58.4	58.6	59.1	59.1	59.7	60.3	61.0	59.4
	(0.6)	(0.6)	(0.6)	(0.6)	(0.6)	(0.6)	(0.6)	(0.6)	(0.6)	(0.2)
(b) Women's singles										
Aces	2.4	2.5	3.4	3.2	3.8	2.7	3.7	3.7	2.1	3.1
	(0.3)	(0.3)	(0.3)	(0.3)	(0.3)	(0.3)	(0.3)	(0.3)	(0.3)	(0.1)
Double faults	6.7	5.4	5.8	5.2	5.6	4.0	5.2	5.4	6.4	5.5
	(0.5)	(0.5)	(0.4)	(0.4)	(0.4)	(0.3)	(0.4)	(0.4)	(0.4)	(0.1)
Service points won	56.2	56.3	55.9	54.8	58.4	56.2	57.7	55.9	53.3	56.1
	(0.9)	(1.0)	(0.8)	(0.8)	(0.8)	(0.8)	(0.9)	(0.9)	(0.9)	(0.3)
1st service in	58.3	61.0	61.3	56.9	61.1	61.9	61.9	63.9	61.4	60.8
	(0.9)	(1.0)	(0.8)	(0.8)	(0.8)	(0.8)	(0.8)	(0.9)	(0.9)	(0.3)

To determine whether serving with new balls is an advantage, let us consider Table 13.5. The age of the balls in games is indicated from 1 (new balls) to 9 (old balls). During the five minutes of warming up before the match begins, the same balls are used as in the first seven games. Thus it makes sense to set the age of the balls in the first game of the match at 3. If the hypothesis that new balls provide an advantage were true, the dominance of service, measured by the probability of winning a point on service, would decrease with the age of the balls. Table 13.5 does not support this hypothesis, at least in the men's singles. For the women the probability of winning a point on service with balls of age 9 is significantly lower than with balls of age 1, but overall there is no evidence for the hypothesis either.

Although serving with new balls appears to provide no advantage in terms of the number of points won, Table 13.5 shows that new balls may affect the *way* that points are won. For example, the probability of "1st service in" seems to increase when the balls become older, and the probability of a double fault seems to decrease, which is, of course, partly due to the increasing trend in the probability of "1st service in." The reason for this is that older balls are softer and fluffier (hence have more grip) than newer balls. The service is therefore easier to control, resulting in a higher percentage of "1st service in" and fewer double faults.

Both effects would result in a greater dominance of service as balls become older. To show why, nevertheless, the dominance of service appears to be independent of

TABLE 13.6
Service characteristics depending on the age of the balls ("age"): logit
estimation results.

Probability	Men's singles		Women's singles	
	constant	age	constant	age
Point won on service	0.580	0.003	0.269	−0.005
	(0.019)	(0.003)	(0.027)	(0.005)
Point won on 1st service	−0.293	0.007 †	−0.593	0.019†
	(0.018)	(0.003)	(0.027)	(0.005)
Point won if 1st service in	1.001	0.002	0.443	0.011
	(0.027)	(0.005)	(0.035)	(0.006)
1st service in	0.341	0.008†	0.340	0.020†
	(0.019)	(0.003)	(0.027)	(0.005)
Point won on 2nd service	0.057	−0.001	0.041	−0.036†
	(0.029)	(0.005)	(0.042)	(0.008)
Point won if 2nd service in	0.414	−0.006	0.378	−0.043†
	(0.031)	(0.006)	(0.046)	(0.008)
2nd service in	1.766	0.017†	1.820	−0.001
	(0.041)	(0.008)	(0.061)	(0.011)

†: Estimate significantly different from zero.

the age of the balls, we split the probability of winning a point on service as follows:

Pr (point won on service)

$\quad =$ Pr (point won on 1st service)

$\qquad +$ Pr (1st service fault) Pr (point won on 2nd service)

$\quad =$ Pr (point won if 1st service in) Pr (1st service in)

$\qquad + \{1 -$ Pr (1st service in)$\}$ Pr (point won if 2nd service in)

$\qquad \times$ Pr (2nd service in).   (13.3)

To analyze how these probabilities depend on the age of the balls, we specify a simple
logit model with a linear function of the age of the balls as the systematic part; see
McFadden (1984). For example, the probability of winning a point on service is
specified as

$$\text{Pr (point won on service)} = \Lambda(\beta_0 + \beta_1 \times \text{age of balls}),\qquad (13.4)$$

where Λ is the logistic distribution function, $\Lambda(x) := \exp(x)/\{1 + \exp(x)\}$. Table 13.6 presents the maximum likelihood estimation results for all probabilities in
Equation (13.3).

As already suggested by Table 13.5, the probability of "1st service in" increases
when balls become older. One might argue that this positive effect on the probability of winning a point on the first service is counteracted by a benefit for the

receiver when balls become older and thus softer and fluffier. The first service would be slower and hence easier to return. We find no evidence for this, as age has no effect on the probability of winning a point if the first service is in. Therefore, in total, players win more points on their first service as balls become older.

The second service is different. The men miss fewer second services when using old balls, which is in line with the decreasing double fault statistics in Table 13.5. However, if the second service is in, they win fewer points, but not significantly so. On balance, the quality of the second service, measured by the probability of winning a point on the second service, is independent of the age of the balls.

For the women the quality of the second service *does* depend on the age of the balls. The second service is easier to return with older balls, which makes the quality of the second service depend negatively on the age of the balls.

Equation (13.3) can now be used to show that, on balance, the age of the balls does not affect the dominance of service. It is true that older balls lead to more points won on the first service. However, both men and women have fewer opportunities to score points on their second service, as the probability of missing a first service decreases. Moreover, the women score fewer points on their second service. On balance, the effects on the first and second service offset each other, so the age of the balls does not affect the quality of service.

A second interpretation of the question whether serving with new balls provides an advantage is that newer balls may benefit the server only in the first game that they are used (age = 1). It may be the transition from old, soft and fluffy balls to new, hard balls that is difficult to cope with for the receiver and/or server. To analyze this we add a dummy variable of balls of age 1 to the logit models used above. However, there is no evidence for an effect of this dummy on the probabilities in Table 13.6. Only for the probability of "2nd service in" for the womens singles does the dummy have a significantly negative effect. This is in line with the high percentage of double faults with new balls in Table 13.5. Including the dummy does not change the effect of the age variable essentially.

In summary, we have looked at three commonly held beliefs, namely that a player is as good as his/her second service, and that serving first in a set or serving with new balls provides an advantage. All three are myths and have no statistical foundation.

13.4   WINNING MOOD

A big issue in the statistical analysis of sports is whether there exists a "winning mood." This relates to the question whether subsequent points are dependent (not whether they are identically distributed; this is taken up in Section 13.5). In this section we consider four commonly heard statements that relate to the possible existence of a winning mood in tennis, first two statements concerning the final set and then two statements regarding breaks.

13.4.1   AT THE BEGINNING OF A FINAL SET, BOTH PLAYERS HAVE THE SAME
         CHANCE OF WINNING THE MATCH

The "final" set—the fifth set in the men's singles (at least in a "Grand Slam" event) and the third set in the women's singles—decides a tennis match. Such a deciding set occurs in about one-fourth of all matches at Wimbledon. Tension is high and mistakes can be costly. There are a number of interesting questions related to the final set. For example, suppose that a "seed" plays against a "nonseed." You wish to forecast the winner. At the beginning of the match, if no further information is available, the relative frequency at Wimbledon that the nonseed wins is small. What is the probability at the beginning of the final set? Is it now close to 50%? Also, is it more difficult for unseeded women to beat a seed than for men, and are men more equal in quality than women? Do players get tired, i.e., does the dominance of the service decrease in long matches (i.e., in the final set). Finally, is it true that, in the final set, the player who has won the previous set has the advantage? Some of these questions will be investigated in this section.

Suppose that a seed plays against a nonseed. At the beginning of the match, if no further information is available, the probability that the nonseed will win is small: 13.1% (2.7%) in the men's singles and 10.5% (2.5%) in the women's singles. (The figures in parentheses are the standard errors.) What is the probability at the beginning of the final set? Some people claim that at the beginning of a final set, both players have the same chance of winning the match.

This is not true. Naturally, the probabilities have increased, but only to 28.7% (8.7%) for the men and 17.1% (6.3%) for the women. Both are significantly different from 50%. The reader may argue that these estimates are biased upwards because of a selection effect (a player ranked 17 is more likely to play and win in a final set against one of the top-16 players, than a player ranked 50), and that this bias is absent in the estimates 13.1% and 10.5% above. The reader would be right, but our conclusion is based on the fact that 28.7% and 17.1% are both significantly smaller than 50%, and this will be even more true if both percentages become smaller after properly accounting for the selection effect. At 2–2 in sets in the men's singles, therefore, it is certainly not true that the chances are now even between the seed and the nonseed. The seed is still very much the favorite. This is even clearer in the women's singles. At 1–1 in sets, the seeded player still has a probability of 82.9% (6.3%) of winning the match!

At first sight, the estimated probabilities of winning the final set — 28.7% for the men and 17.1% for the women — also seem to indicate that it is more difficult for a female nonseed to beat a seed than is the case for a man, as is often claimed. However, one should keep in mind that the male nonseed has to win two sets to arrive at the final set, whereas the woman has to win only one set. (The fact that the tennis scoring system has an impact on the probability of winning a match was investigated in Jackson, 1989; Maisel, 1966; Miles, 1984; Riddle, 1988; and Riddle, 1989.) Therefore, when arriving at the final set, the quality difference between the nonseed and the seed is generally smaller for the men than for the women. This in itself will result in a higher probability of winning the final set for unseeded men than for unseeded women.

13.4.2 IN THE FINAL SET THE PLAYER WHO HAS WON THE PREVIOUS SET HAS THE ADVANTAGE

We next consider the relation between the final and "prefinal" sets. One often hears that in the final set, the player who has won the previous set has the advantage. Is this true? We investigate this hypothesis by looking at probabilities of winning the final

TABLE 13.7
Estimated probabilities of winning fifth (third) set after winning fourth (second) set in men's (women's) singles.

	Probability (%)				
	Sd-Sd	Sd-NSd	NSd-Sd	NSd-NSd	Total
Men's singles	10.7	40.8	76.4	57.6	50.2
	(10.3)	(11.3)	(15.0)	(12.8)	(7.0)
Women's singles	33.7	62.3	14.8	68.5	61.2
	(15.8)	(8.9)	(14.5)	(13.4)	(6.5)

set after winning the "prefinal" set. These probabilities are presented in Table 13.7 for all matches taken together (Total) and for different types of matches. Sd-NSd indicates again a match between a seeded player and an unseeded player, where the first player (Sd) won the "prefinal" set. Sd-Sd, NSd-Sd, and NSd-NSd are similarly defined. Because the number of observations is quite small (51 final sets in the men's singles and 57 final sets in the women's singles in our data set), the standard errors are quite large.

In the men's singles, the probability that the same player wins the fourth and fifth sets is estimated to be 50.2% (7.0%). In the women's singles, the estimated probability that the same player wins the second and third sets is 61.2% (6.5%). These percentages are not significantly different from 50%, so there appears to be no basis for the hypothesis in question.

If we look at the subcategories, then we see that, when two seeds play against each other, the winner of the fourth (second) set will probably lose the match, especially in the men's singles. When an unseeded woman plays against a seed, winning the "prefinal" set is also a disadvantage, because her probability of winning the match is only 14.8% (14.5%). Both results are significant and indicate that, if there is correlation between the final and "prefinal" sets, then this is more likely to be negative than positive.

13.4.3 AFTER BREAKING YOUR OPPONENT'S SERVICE THERE IS AN INCREASED
CHANCE THAT YOU WILL LOSE YOUR OWN SERVICE.

The server is expected to win his or her service game, particularly on fast surfaces such as grass. If he or she fails to do so, this is considered serious in women's singles and disastrous in men's singles. A game not won by the server but by the receiver is called a "break."

As we show in Klaassen and Magnus (2001), *points* in a tennis match are not independent. Maybe the same holds at game level. If so, this dependence of games would have important consequences for the statistical modeling of tennis matches. We now ask whether games are independent, and, if so, where this comes from. One source of dependence of games could be the often heard "break-rebreak" effect: After breaking your opponent's service there is an increased chance that you will lose your own service.

At first sight there appears to be no support for the "break-rebreak" effect in Table 13.8. Overall, the probability of winning a service game increases rather than decreases after a break, in the men's singles from 80.1% to 83.4%, in the women's singles from 61.2% to 66.9%. The increase is 3.3%-points (1.1%) for the men and 5.7%-points (1.6%) for the women, hence both are significant. The reader may protest by

TABLE 13.8
Estimated probabilities of winning a service game depending on what happened in the previous game (same set).

	Previous game	Sd-Sd	Sd-NSd	NSd-Sd	NSd-NSd	Total
Men	Break	86.7	87.0	73.5	83.5	83.4
		(3.0)	(1.4)	(2.6)	(1.8)	(1.0)
	No break	85.6	89.4	74.3	78.6	80.1
		(1.2)	(0.7)	(0.9)	(0.9)	(0.5)
Women	Break	67.5	77.7	50.6	65.6	66.9
		(3.5)	(1.7)	(2.8)	(2.9)	(1.3)
	No break	65.4	77.8	48.3	61.2	61.2
		(2.5)	(1.6)	(1.6)	(2.2)	(1.0)

arguing that we should only consider matches between players of equal strength, Sd-Sd or NSd-NSd, because with players of unequal strength (Sd-NSd and NSd-Sd) it is to be expected that the stronger player often breaks the weaker player and then goes on to win his/her own service game, whether a "break-rebreak" effect exists or not. However, even if we constrain ourselves to matches between players of equal strength, then the "break-rebreak" effect is still absent, as Table 13.8 shows. Hence, our conclusion is unchanged: after a break, the players are more, rather than less, likely to win their service game. Thus, in the game following a break, it is not true that the winner takes it a bit easier and the loser tries a bit harder. Apparently, what happens is just the opposite. The winner gains in confidence and the loser gets discouraged. So there is positive correlation across games. Note that this effect is stronger in NSd-NSd matches than in Sd-Sd matches. Seeds appear to play less dependently than nonseeds.

13.4.4 AFTER MISSING BREAK POINTS IN THE PREVIOUS GAME THERE IS AN INCREASED CHANCE THAT YOU WILL LOSE YOUR OWN SERVICE

Suppose now that in the previous game no break has occurred, but that the receiver had a good chance to break because he/she had one or more break points. The receiver did not, however, capitalize on these break points and this may have discouraged him/her. Does such an effect occur, and does it affect the current game? That is, is it true that after missing one or more break points in the previous game there is an increased chance that you will lose your own service? Table 13.9 shows no support

TABLE 13.9
Estimated probabilities of winning a service game depending on what happened in the previous game (same set).

	Previous game	Sd-Sd	Sd-NSd	NSd-Sd	NSd-NSd	Total
Men	Break point(s),	87.7	91.8	70.5	76.3	79.0
	but no break	(3.1)	(1.6)	(2.8)	(2.6)	(1.3)
	No break points	85.3	89.0	74.8	78.9	80.2
		(1.3)	(0.8)	(1.0)	(1.0)	(0.5)
Women	Break point(s),	61.3	75.2	48.7	50.9	56.9
	but no break	(5.6)	(3.4)	(4.3)	(5.6)	(2.4)
	No break points	66.4	78.5	48.3	62.8	61.9
		(2.7)	(1.8)	(1.7)	(2.4)	(1.1)

for this idea for the men, as the probability of a break is only slightly higher than usual: 1.2%-points (1.4%). For the women, however, the increase is 5.0%-points (2.6%), so missing break points seems to affect the next game. Women play the games more dependently than men. More specifically, there is a positive correlation across games: discouragement because of missed break points causes problems in the next game.

The "discouragement effect" is particularly strong when two nonseeds play against each other. There the probability of losing a service game after missed break points in the previous game increases by 11.9%-points (6.1%) in the women's singles. This adds support to the idea that seeded players are not only technically but also mentally stronger than nonseeded players. They don't let themselves be discouraged and are less affected by what happened in the previous game. This is also supported by the men's percentages.

Hence, *games* in a tennis match appear not to be independent. More specifically, there is a positive correlation across games: if you did well in the previous game, you will do well in the current game. For example, after breaking your opponent's service, you are in a "winning mood" and are more likely to win your own service game. Moreover, if you had break points in the previous game but missed them all, it is more difficult to win your own service game, at least for the women, not for the men. The correlation across games is more present in matches between two nonseeds than in matches between two seeds, thus supporting the idea that seeded players are not only technically but also mentally stronger than nonseeds. These results at game level correspond to the positive correlation at point level found in Klaassen and Magnus (2001).

13.5   BIG POINTS

Few readers would disagree that a point played at 30–40 in a tennis game is more important than, say, at 0–0. Similarly, a game at 5–4 in a set is more important than a game at 1–0. And the final set is more important than the first set. The question which concerns us here is whether players play each point "as it comes" or whether they are affected by what happened in the previous point (ace, double fault) and/or by the importance of the point (break point, game point). Do players make sure their first service is in at break point or game point? Do the "real champions" play their best tennis at the big (important) points, or do big points not exist? Do "big games" exist, for example the seventh game? If there are big points and if players do not play each point as it comes, then clearly points are not identically distributed and this will have important consequences for the statistical modeling of tennis matches (see also Klaassen and Magnus, 2001).

13.5.1   THE SEVENTH GAME

Our first myth is not about big points but about big games, and it is a commentators' favorite—the idea that the seventh game is especially important. We don't know the history of this "wisdom," but there is no doubt that many believe it. We can measure the importance of a game in two ways, unweighted or weighted. The unweighted importance of a game in a set is defined as the probability that the server in that game wins the set given that he/she wins the game minus the probability that the server wins the set given that he/she loses the game. Note that if we had defined the unweighted importance of a game in a set as the probability that the server in that game wins the set given that he/she wins the game (so not minus the other probability), then we would not have taken account of the fact that, after the first set, in odd games usually the player who won the previous set (often the weaker player) serves. This would have biased the importance measure. The unweighted importance does not take into account that some games (like the 12th) occur much less frequently than other games (such as the first 6, which occur in every set). The weighted importance takes this into account, since the weighted importance is defined as the unweighted importance times the probability that the game occurs.

TABLE 13.10

(Un)Weighted importance of each game in a set.

Game	Men's singles Unweighted	Weighted	Women's singles Unweighted	Weighted
1	36.4 (3.5)	36.4 (3.5)	27.1 (4.5)	27.1 (4.5)
2	31.9 (3.7)	31.9 (3.7)	42.5 (4.2)	42.5 (4.2)
3	40.4 (3.2)	40.4 (3.2)	39.0 (4.1)	39.0 (4.1)
4	40.8 (3.5)	40.8 (3.5)	40.0 (4.2)	40.0 (4.2)
5	35.9 (3.5)	35.9 (3.5)	38.1 (4.3)	38.1 (4.3)
6	44.0 (3.4)	44.0 (3.4)	33.1 (4.3)	33.1 (4.3)
7	39.4 (3.3)	38.6 (3.3)	34.9 (4.4)	32.7 (4.2)
8	30.1 (4.3)	26.6 (4.0)	47.4 (4.7)	36.8 (4.0)
9	36.9 (4.0)	27.8 (3.3)	34.2 (5.6)	19.5 (3.7)
10	58.2 (4.1)	31.6 (2.6)	55.4 (6.7)	20.2 (3.5)
11	57.0 (4.6)	15.7 (1.6)	39.8 (10.6)	6.0 (2.4)
12	67.7 (4.4)	18.9 (1.6)	67.1 (7.9)	10.0 (1.9)
Mean	43.2 (1.1)	32.4 (0.9)	41.6 (1.7)	28.8 (1.1)

As Table 13.10 shows, the idea that the 7th game is singled out to be important is nonsense. Unweighted, games 10, 11, and 12 in the men's singles and games 10 and 12 in the women's singles are the most important. There is no indication that the 7th game plays a special role. Since the 10th, 11th, and 12th games don't occur as frequently as the first 6 games, their weighted importance is rather less. Weighted, the 7th game does not play a special role either. Hence, we reject this hypothesis. In addition, Table 13.10 confirms again that losing a service game in the men's singles is more serious than in the women's singles, because the average importance of a game in the men's singles is 32.4 compared to 28.8 in the women's singles.

13.5.2 DO BIG POINTS EXIST?

Let us now look more directly at the existence of big points. Is it true that all points are equally important, or do big points exist?

We can measure the strength of a seeded player relative to a nonseeded player by $p_{Sd} - p_{NSd}$, where p_{Sd} denotes the estimated probability that the seed wins a point while serving against a nonseed, and p_{NSd} denotes the estimated probability that the nonseed wins a point while serving against a seed. For the men, $p_{Sd} = 69.3\%$ and $p_{NSd} = 61.1\%$, whereas for the women $p_{Sd} = 62.9\%$ and $p_{NSd} = 50.1\%$; see Table 13.3. Hence,

$$p_{Sd} - p_{NSd} = \begin{cases} 8.2\% & \text{for the men} \\ 12.8\% & \text{for the women.} \end{cases}$$

The difference between seeds and nonseeds is thus greater in the women's singles than in the men's singles, which is no surprise. There is a second way to measure the difference in strength, namely the estimated probability that the seed wins a set from the nonseed. This is 75.1% in the men's singles and 82.0% in the women's singles and we reach the same conclusion. In Table 13.11 we present the two measures for each set. We see that the difference in strength between a seeded and a nonseeded player (as measured by $p_{Sd} - p_{NSd}$ decreases gradually in the men's singles, but not in the women's singles. However, in both the men's singles and the women's singles, the difference in strength is least in the final set (4.4% in the men's singles, 10.7% in the women's singles). This is what one would expect. In the deciding final set we don't expect a lot of difference in strength any more, even though a seed and a nonseed play each other.

TABLE 13.11

Two measurements of the strength of a seeded player relative to a nonseeded player.

Set	Men's singles		Women's singles	
	$p_{Sd} - p_{NSd}$	Pr(Sd wins set)	$p_{Sd} - p_{NSd}$	Pr(Sd wins set)
1	9.6	79.9	11.7	78.1
	(1.0)	(3.2)	(1.0)	(3.4)
2	8.8	78.2	14.7	85.6
	(1.0)	(3.3)	(1.1)	(2.9)
3	7.6	70.1	10.7	82.9
	(1.0)	(3.6)	(2.1)	(6.3)
4	7.3	70.7	—	—
	(1.4)	(5.3)		
5	4.4	71.3	—	—
	(2.1)	(8.7)		
Total	8.2	75.1	12.8	82.0
	(0.5)	(1.8)	(0.7)	(2.1)

Note: p_{Sd}: estimated probability of seeded player winning a point on service versus a nonseeded player; p_{NSd}: estimated probability of nonseeded player winning a point on service versus a seeded player.

Now a curious phenomenon occurs. Even though the relative number of points won by the seed is lowest in the final set, this is not true for the probability that the seed wins the set. The seed still has a large chance to win the final set. And this is true both for the men and for the women; see Section 13.4.1. There is only one possible explanation, and that is that some points are more important than others and that the seeds play the important points better (or equivalently, that the nonseeds play them worse). Thus, big points do exist.

13.5.3 REAL CHAMPIONS

We have just seen that seeds play the big points in the final set better than nonseeds. Is this true for the whole match? Many people believe this, as we often hear that real champions play their best tennis at the big points.

Let us think of all break points as big points, a drastic simplification. When a seed serves in the men's singles, the probability of winning a point on service is 68.3% (1.2%) at break point down and 68.7% (0.3%) at other points (seeded and nonseeded receiver taken together). The difference is not significant. However, when a nonseed serves, the difference is significant: 59.2% (0.9%) at break point and 63% (0.3%) at other points. Therefore, real champions perform better at break point down than other players. However, real champions do not win more points at break point down. It is the nonseeds who win fewer points at break point down. So, it seems that the seeded players do not play their best tennis, but rather that the nonseeded players play their worst. Another interpretation is that the seeded server, break point down, performs better and so does the receiver (seeded or nonseeded), and that these two improvements cancel each other. In contrast, a nonseeded server, break point down, does not perform better, while the receiver (seeded or nonseeded) does.

In contrast to the men, we find no evidence for our hypothesis for the women. In the women's singles, the probability of winning a point on service is 58.2% (1.4%) at break point down and 61.6% (0.4%) at other points for a seed (seeded and nonseeded receiver taken together). For a nonseed these are 50.8% (1.1%) at break point down and 54.9% (0.4%) otherwise. Both differences are significantly negative, and the decrease for the seeds (3.4%-points (0.6%)) is only marginally smaller than that of the nonseeds (4.1%-points (1.2%)). So real champions do not perform better than others in the women's singles.

The word "real champion" can also be interpreted in a relative, rather than an absolute, manner. In an absolute sense, a player is a real champion if he/she is very good, for example a seed, irrespective of the opponent. In a relative sense, a player is a real champion if he/she performs particularly well at important points depending on the opponent.

To analyze whether (relative) real champions play their best tennis at break points, we use Table 13.12. The differences in this table show the relative service quality at break point down with respect to other points. If the hypothesis were true, a server would perform better against a nonseed than against a seed using the differences as quality measures. For the men we see that a seed performs relatively better against a nonseed than against a seed, but the difference of 1.7 percentage points ($0.1 - -1.6$) is insignificant. However, a nonseed plays relatively worse against a nonseed than against a seed (again insignificantly). So we find no evidence for our hypothesis in relative sense for the men. The same holds for the women.

Let us now compare the results for absolute and relative champions with the conclusion at the end of the discussion of the hypothesis about the existence of big points. Table 13.11 concerns only matches between a seed and a nonseed. So, a distinction between absolute and relative real champions is no longer possible, since seeds are champions both in absolute and relative sense compared to nonseeds.

TABLE 13.12
Estimated probabilities of winning a point on service.

		Sd-Sd	Sd-NSd	NSd-Sd	NSd-NSd	Total
Men	At break point	65.5	69.4	57.8	59.8	60.9
		(2.2)	(1.4)	(1.2)	(1.3)	(0.7)
	At other points	67.1	69.3	61.4	64.1	64.7
		(0.6)	(0.4)	(0.4)	(0.4)	(0.2)
	Difference	−1.6	0.1	−3.6	−4.3	−3.8
		(2.3)	(1.5)	(1.3)	(1.4)	(0.7)
Women	At break point	58.2	58.2	48.8	51.6	52.3
		(2.2)	(1.7)	(1.3)	(1.9)	(0.9)
	At other points	56.7	63.4	50.3	56.4	56.6
		(0.8)	(0.5)	(0.5)	(0.7)	(0.3)
	Difference	1.5	−5.2	−1.5	−4.8	−4.3
		(2.3)	(1.8)	(1.4)	(2.0)	(0.9)

Therefore, we aggregate the results regarding our hypothesis across absolute and relative champions. Then our conclusion is that male seeds play break points better than nonseeds. This is in line with the conclusion from Table 13.11 that seeds play the important points in the final set better than nonseeds. Therefore, our overall conclusion for the men is that our hypothesis is true. For the women we did not find evidence for our hypothesis in terms of absolute and relative champions. However, Table 13.11 shows that seeds play the key points in the third set better than nonseeds. So, our overall conclusion for the women is that our hypothesis might be true.

13.6   CONCLUSION

The television commentator at a tennis match is not to be envied. Repeating the names of the players (as football commentators do) is hardly useful. The score is typically presented in the screen and mentioning it is thus superfluous. So, what must they talk about? A match at Wimbledon in the men's singles lasts, say, two-and-a-half hours. The effective time-in-play for one point is about five seconds. With six points per game, a game lasts thirty seconds. With ten games per set, a set lasts five minutes. With four sets per match, a match lasts only twenty minutes. The remaining two hours and ten minutes are to be filled by the commentators. It follows that they play a major part in the broadcast. But are the things that they say correct? Are the idées reçues of the commentators true? Unfortunately, as we have found out in this chapter, almost none of them are.

ACKNOWLEDGEMENTS

We are grateful to the Royal Statistical Society and to Taylor and Francis Ltd. for permission to use material published earlier by us in *The Statistician* (volume 48, Part

2, pp. 239–246 and 247–256, 1999) and the *Journal of Applied Statistics* (volume 26(4), pp. 461–468, 1999). As always we thank IBM UK and The All England Club at Wimbledon for their kindness in providing the data.

REFERENCES

Anderson, C.L. (1977). Note on the advantage of first serve. *Journal of Combinatorial Theory A 23*, 363.

Borghans, L. (1995). Keuzeprobleem op Centre Court. *Economisch Statistische Berichten 80*, 658–661.

Gale, D. (1980). Optimal strategy for serving in tennis. *Mathematics Magazine 44*, 197–199.

George, S.L. (1973). Optimal strategy in tennis: A simple probabilistic model. *Applied Statistics 22*, 97–104.

Gillman, L. (1985). Missing more serves may win more points. *Mathematics Magazine 58*, 222–224.

Hannan, E.L. (1976). An analysis of different serving strategies in tennis. In R.E. Machol, S.P. Ladany, and D.G. Morrison (Eds.), *Management Science in Sports*, pp. 125–135. New York: North-Holland.

Jackson, D.A. (1989). Letter to the editor on "Probability models for tennis scoring systems" by L.H. Riddle. *Applied Statistics 38*, 377–378.

Kingston, J.G. (1976). Comparison of scoring systems in two-sided competitions. *Journal of Combinatorial Theory 20*, 357–362.

Klaassen, F.J.G.M. and J.R. Magnus (2001). Are points in tennis independent and identically distributed? Evidence from a dynamic binary panel data model. *Journal of the American Statistical Association 96*, 500–509.

Klaassen, F.J.G.M. and J.R. Magnus (2006). Are economic agents successful maximizers? An analysis through service strategy in tennis. Discussion Paper 2006-52, CentER, University of Tilburg, Tilburg, The Netherlands. Submitted for publication.

Little, A. (1995). *Wimbledon Compendium 1995*. London: All England Lawn Tennis and Croquet Club.

Magnus, J.R. and F.J.G.M. Klaassen (1999a). The effect of new balls in tennis: Four years at Wimbledon. *The Statistician 48*, 239–246.

Magnus, J.R. and F.J.G.M. Klaassen (1999b). The final set in a tennis match: Four years at Wimbledon. *Journal of Applied Statistics 26*, 461–468.

Magnus, J.R. and F.J.G.M. Klaassen (1999c). On the advantage of serving first in a tennis set: Four years at Wimbledon. *The Statistician 48*, 247–256.

Maisel, H. (1966). Best k of $2k - 1$ comparison. *Journal of the American Statistical Association 61*, 329–344.

McFadden, D. (1984). Econometric analysis of qualitative choice models. In Z. Griliches and M.D. Intriligator (Eds.), *Handbook of Econometrics*, Volume II, Chapter 24. Amsterdam: North-Holland.

Miles, R.E. (1984). Symmetric sequential analysis: The efficiencies of sports scoring systems (with particular reference to those of tennis). *Journal of the Royal Statistical Society, Series B 46*, 93–108.

Norman, J.M. (1985). Dynamic programming in tennis — when to use a fast serve. *Journal of Operational Research Society 36*, 75–77.

Riddle, L.H. (1988). Probability models for tennis scoring systems. *Applied Statistics 37*, 63–75 and 490.

Riddle, L.H. (1989). Author's reply to D.A. Jackson. *Applied Statistics 38*, 378–379.

14

Back to back evaluations on the gridiron

David J. Berri

California State University-Bakersfield

ABSTRACT

Economists have utilized player performance statistics from both baseball and basketball to measure a player's marginal physical product. Such information can then be used to examine various issues in labor economics. American football also offers various statistics designed to measure player performance. This essay will show that these performance statistics can be connected to team outcomes on the gridiron. Furthermore, we will explore how well these statistics ultimately translate into a measure of marginal physical product for quarterbacks and running backs in the sport of football.

14.1 WHY DO PROFESSIONAL TEAM SPORTS TRACK PLAYER STATISTICS?

In The Numbers Game, Alan Schwartz details the history of player performance measures in professional baseball. Soon after players took the field for professional teams in the 19th century people began to track such measures as hits and at-bats. And in 1870, H.A. Dobson noted that a baseball player's efficiency could be measured simply with batting average — calculated simply by dividing hits by at-bats.

As time went by increasingly sophisticated metrics were developed, such as on-base percentage, slugging percentage, *OPS* (on-base percentage plus slugging average), among others. Tracking the statistics and developing new metrics does require the expenditure of some effort. Is there a benefit generated to offset this cost?

The answer to this query begins with why player statistics are tracked in the first place. For any game we can see which team won and which team lost by looking at the scoreboard. The question we have is which players on each team were primarily responsible for the outcome we observe.

To get at this question, interested observers, i.e., teams, the media, fans, analyze player statistics. The purpose behind this effort is to separate each player from his team and connect that specific player's actions accurately to the team outcome observed.

There are two reasons why we wish to separate the player from his team. First, we wish to explain why a specific team has won or lost. Specifically, which players are most responsible for the outcome we observed? From the team's perspective, though, there is a more important issue. Teams need to know which players to employ in the future. By tracking and analyzing player statistics teams hope to identify the players who will help the team be successful in the future. In sum, statistics are tracked to both explain what we observed in the past and determine what actions a team should take in the future.

In baseball the first task is relatively easy. The statistics tracked for baseball players — singles, doubles, triples, home runs, etc. — are clearly linked to runs scored and wins. And understanding the value of these statistics is fairly easy. One does not need advanced regression analysis to understand that one more home run is worth more to a team than one more single.

As noted in Berri, Schmidt, and Brook (2006), baseball performance is not entirely consistent across time. So forecasting future performance in baseball is somewhat difficult, even if one completely understands the data collected.

Basketball also has an abundance of player statistics. Teams track for each player points, rebounds, steals, assists, and other factors to help decision-makers both explain and predict performance. Relative to baseball, the explanation of why teams win and lose is a bit more difficult in basketball. After all, which is more important, an additional point scored or an assist? Would a team rather have one more rebound or one more blocked shot?

As noted in Berri et al. (2006), with a bit of thought one can untangle the relative value of these statistics. Furthermore, relative to baseball, players in basketball tend to be more consistent across time. Hence, player forecasts are relatively more reliable in hoops.

What about the American football? Yes, a number of statistics are tracked. But statistics are only commonly tracked for certain positions. Quarterbacks and running backs each have a number of factors tracked and these clearly can be used to assess performance. Other positions, though, such as offensive lineman, have hardly any statistics. Still, numbers do exist in football. The purpose of this essay is to explore how these statistics can be used to explain past performance, as well as predict the future. In other words, do statistics in the National Football League (NFL) have the same value as statistics tracked in baseball and basketball?

14.2   THE NFL'S QUARTERBACK RATING MEASURE

For quarterbacks the NFL tracks a variety of statistics, including pass attempts, completions, yards passing, interceptions thrown, touchdown passes, rushing attempts, rushing yards, rushing touchdowns, yards lost from sacks, the number of times sacked, fumbles, and fumbles lost. In evaluating quarterbacks the data on rushing, sacks, and fumbles tends to be minimized. This can be seen when one considers the NFL's widely accepted quarterback rating measure, which is perhaps the most complicated metric in all of professional sports.

To understand this statement, consider the following step-by-step description of this measure as it was reported in Berri et al. (2006).

"First one takes a quarterback's completion percentage, then subtracts 0.3 from this number and divides by 0.2. You then take yards per attempt, subtract 3 and divide by 4. After that, you divide touchdowns per attempt by 0.05. For interceptions per attempt, you start with 0.095, subtract from this number interceptions per attempt, and then divide this result by 0.04. To get the quarterback rating, you add the values created from your first four steps, multiply this sum by 100, and divide the result by 6. Oh, and by the way, the sum from each of your first four steps cannot exceed 2.375 or be less than zero."

It does not become any clearer when one considers the mathematical formulation of the NFL's quarterback rating.

$$\left(\frac{\frac{COMP}{PASSATT} - 0.3}{0.2} + \frac{\frac{PASSYDS}{PASSATT} - 3}{4} + \frac{\frac{PASSTD}{PASSATT}}{0.05} + \frac{\frac{0.095 - INT}{PASSATT}}{0.04} \right) \times \frac{100}{6}.$$

Where $COMP$ = Completions, $PASSYDS$ = Yards passing, $PASSTD$ = Touchdown passes thrown, INT = Interceptions thrown, and $PASSATT$ = Passing attempts. The NFL's quarterback rating system is featured prominently in broadcasts of each NFL game and at NFL.com. Is this an accurate reflection of a quarterback's value? In addition to being both complex and nonintuitive, the measure also ignores rushing attempts, rushing yards, sacks, yards lost from sacks, and fumbles lost. Furthermore it is not clear that the factors that are included are actually weighted correctly in terms of team outcomes. In sum, this measure is quite complex, clearly incomplete, and of questionable accuracy. Is it possible to do better?

14.3 THE SCULLY APPROACH

To answer this question we turn to the classic work of Scully (1974). Scully examined the rate of monopsonistic exploitation in baseball prior to the era of free agency. An investigation of exploitation requires the researcher to ascertain a worker's marginal revenue product, a step that requires a measure of worker productivity. To accomplish this objective, Scully proposed a simple model of team wins, where a team's winning percentage was regressed on, among other factors, a team's slugging percentage and its strike-out to walk ratio. From this regression Scully was able to determine the contribution each player made to wins, which is in essence his marginal physical product.

Blass (1992) improved upon the Scully approach with respect to hitters in baseball. Borrowing from methods advanced by the Society for the Advancement of Baseball Research (SABR), and echoing specifically the work of Thorn and Palmer (1984), Blass regressed runs scored on a host of statistics tracked for the individual batter. This list included factors incorporated in slugging percentage, but also such factors as stolen bases and walks. From this model Blass was able to ascertain the value — in terms of runs scored — of a host of statistics tracked for the individual hitter in baseball.

More recently the Scully-Blass approach has been applied to basketball. In Berri et al. (2006) a measure of each player's Wins Produced is described.* Like the earlier work of Scully and Blass, the Wins Produced model involves regressing an outcome — in this case team winning percentage — on team offensive and defensive efficiency. Given that the two efficiency measures incorporate a variety of statistics tracked for individual players, the model of winning percentage allows one to determine the value of each player's statistical production.

Like baseball and basketball, football also tracks various measures of player performance. One might wonder if these statistics can be translated into a measure of each player's marginal physical product that improves upon the NFL's quarterback rating system. In Berri et al. (2006) such a model was offered. What follows are the details of this effort.

14.4   MODELING TEAM OFFENSE AND DEFENSE

Applying the Scully-Blass approach to football begins with first identifying the appropriate team outcome. Like baseball, football is neatly divided into offense and defense. Consequently, following the lead of Blass, the outcome variable will not be team wins, but points scored. Specifically, two models will be developed, the first describing how many points a team's offense creates. The second will model the number of points a team's defensive allows.

To model offensive points one must first specify the correct dependent variable. A team can score points via its offense, defense, or special teams. In modeling offensive points (*OFFPTS*), though, points drawn from defense and special teams cannot be employed in the measurement of the dependent variable.

Given the focus on offensive point production, *OFFPTS* is calculated by first noting that a team's offense can score via touchdowns from its rushing attack (*RUSHTD*) or its passing game (*PASSTD*). For each touchdown a team has the opportunity to score either one or two extra points (*XP*). If a team fails to score a touchdown, a team can also score points via field goals (*FGM*). This is all summarized in equation (14.1).

$$OFFPF = 6 \times RUSHTD + 6 \times PASSTD + n \times XP + 3 \times FGM. \qquad (14.1)$$

Equation (14.1) has one unknown. The NFL does not record how many extra points are derived from offensive touchdowns and how many come from touchdowns generated by a team's special teams or defense. To ascertain the value of (*n*), one can look at the percentage of touchdowns scored by the team's offense and simply assume that this percentage represents the percentage of extra points scored by the team's offense.

With our dependent variable constructed, we can now construct a list of relevant explanatory variables. As noted in our discussion of the NFL's quarterback rating,

*The model described in Berri et al. (2006) was introduced in Berri (2004) — and referenced in Berri and Schmidt (2006), Lee and Berri (forthcoming), Berri and Krautmann (2006), Berri and Eschker (2005), Berri, Brook, Fenn, Frick, and Vicente-Mayoral (2005), and Berri and Fenn (2004).

a signal caller can be evaluated in terms of rushing, passing, interceptions thrown, and fumbles lost. Consequently, one might expect us to simply regress our measure of offensive scoring, *OFFPTS*, upon the statistics tabulated for the quarterbacks or running backs. Such an approach, though, does not control for all the other factors that impact scoring. The issue here is that the quarterback and running backs are only partially responsible for offensive scoring. To assess each player's contribution we must first identify the complete list of factors that explain offensive scoring. When this task is completed, we can tease out the portion which we can credit to the quarterback or running back.

Our explanation of offensive scoring is presented in Table 14.1. The variables may be separated out into four basic categories: acquiring the ball, moving the ball, keeping the ball, and scoring. The story goes a little like this: For a team to score they first need to have possession of the ball. Once they have possession, they need to move the ball down the field. Moving the ball is not enough as the team must hold on to the ball, i.e., make first downs, if it wishes to keep possession. Finally, if the team can keep moving with the ball, the team has to actually score. Table 14.1 describes in detail the statistics that are used to capture each of these steps. To ease the discussion, we will examine each category of statistics separately.

CATEGORY 1 - ACQUIRING THE BALL- OFFENSE

We begin with a team acquiring the ball. A team can gain possession of the ball via the opponent's kick-off (*DKO*),[†] punts (*DPUNTS*), and also missed field goals (*DFGMISS*). Generally, each of these actions gives the team possession within a team's half of the field, forcing the team to gain substantial yardage to score. A team can also gain possession when the opponent turns the ball over via an interception (*DINT*) or a lost fumble (*DFUMLST*). These events can give a team possession in various places on the field, including deep in the opponent's half of the field. Hence we expect interceptions and fumbles to have a greater positive impact on scoring.[‡]

CATEGORY 2 - MOVING THE BALL- OFFENSE

Once a team gains possession of the ball, the team must move the ball down the field. In Berri et al. (2006) the model incorporated factors that indicated the outcome of the play that gave the team possession. Hence, the average yardage returned on kick-offs (*KORETAVG*), punts (*PUNTRETAVG*), and interceptions (*INTRETAVG*) was incorporated. In addition, the average yardage on the opponent's punts (*DPUNTAVG*) was also employed as an independent variable. An alternative to these four variables is to

---

[†]The primary source of data employed did not report the number of opponent kick-offs. Given that kick-offs occur after a touchdown, a field goal, and at either the start of the game or the start of the second-half, this variable was simply estimated. There was a source that reported this factor for the 2004 season. The correlation between the predicated number of kick-offs, employed here, and the actual number recorded for 2004 was 0.98, suggesting the measure utilized here was a reasonable approximation.

[‡]A team also can gain possession of the ball when the opponent turns the ball over on downs. The source of data (NFL Record & Fact Book), did not record this event.

TABLE 14.1
Factors impacting a team's offensive ability.

Actions	Variables tabulated
Acquisition of the ball	Opponent's kick-offs (*DKO*) Opponent's punts (*DPUNTS*) Opponent's missed field goals (*DFGMISS*) Opponent's interceptions (*DINT*) Opponent's fumbles lost (*DFUMLST*)
Moving the ball	Average starting position of drives (*START*) Total offensive yards gained $OFFYDS =$ $RUSHYDS + PASSYDS$ Total rushing yards gained (*RUSHYDS*) Total passing yards gained (*PASSYDS*) Total penalty yards lost (*PENYDS*) Total penalty yards lost by the opponent (*DPENYDS*)
Maintaining possession	$PLAYS = RUSHATT + PASSATT + SACKED$ Rushing attempts (*RUSHATT*) Passing attempts (*PASSATT*) Sacks (*SACKED*) Third down conversion rate (*3RDCON*) Missed field goals (*FGMISS*) Interceptions (*INT*) Fumbles lost (*FUMLOST*)

Scoring	Touchdown rate $TDRATE =$
	      $OFFTD/(OFFTD + FGMADE)$
	 Touchdowns scored by a team's offense (*OFFTD*)
	 Field goals made (*FGMADE*)
	 Extra points conversion rate $XPRATE$
	      $= OFFXP/OFFTD$
	 Extra points earned on offensive touchdowns (*OFFXP*)

simply note that average starting position of each team's offensive drives (*START*), a factor reported by Football Outsiders.com. The inclusion of this factor does improve the explanatory power of the model reported here — relative to the work reported in Berri et al. (2006) — although the ultimate conclusions are unchanged.

Having acquired the ball, the team must now move across the field of play. The obvious measure of this action is the number of yards the team's offense gains (*OFFYDS*). A team can gain yards via running (*RUSHYDS*) and passing the ball (*PASSYDS*). Yardage can also be gained or lost via penalties (*PENYDS, DPENYDS*). Specifically, penalties on the offense will reduce a team's offensive ability while defensive penalties improve a team's offense.

CATEGORY 3 - KEEPING THE BALL - OFFENSE

Of course, yards are not the only story. If a team moves too slowly, it will lose possession of the ball. When a team gains possession of the ball it has four plays, called downs, to travel ten yards and earn a first down. Given the system of downs, the more plays the team runs (*PLAYS*), the lower the team's ability to score. Plays include rushing attempts (*RUSHATT*), passing attempts (*PASSATT*), and sacks (*SACKED*). One would expect that a play, without any yards gained, would reduce a team's offensive output. In other words, *PLAYS* should have a negative coefficient.

Although a team technically has four downs to gain ten yards, the fourth down is rarely employed to advance the ball. If a team tries to move the ball on fourth down and fails to gain a first down, it loses possession of the ball. Hence, fourth down is where a team will generally punt or try for a field goal. In terms of moving the ball, then, it is third down that is of most importance. The ability of a team to gain a first down on third down, measured via its third down conversion rate (*3RDCON*), is also thought to be important in determining offensive success.

As noted, a team will lose possession of the ball when it punts. Given the relationship between the number of punts and many of our remaining variables, this factor should not be included in our analysis. Other factors associated with losing possession of the ball, including missed field goals (*FGMISS*), interceptions thrown (*INT*), and fumbles lost (*FUMLOST*) are employed.

CATEGORY 4 - SCORING - OFFENSE

If a team does not turn the ball back over to its opponent, and time does not expire, it will ultimately score. Of course, scoring via a touchdown is preferred to a field goal. To gauge the effectiveness of a team's offense we need to note how many points a team scored via each method, or its *TDRATE*. This is calculated by simply dividing offensive touchdowns by the sum of offensive touchdowns and field goals made. For example, the 2004 Detroit Lions scored seven rushing touchdowns, nineteen passing touchdowns, and kicked 24 field goals. Consequently the team's *TDRATE* was 52%.

The last factor listed in Table 14.1 returns us to the subject of extra points. Beyond noting how many extra points we can attribute to the offense, we also wish to note how many points a team was able to earn on these plays. A team can kick an extra point, and if successful, earn one point. Or a team can try and run or pass for two points. Failure to convert these tries results in zero points. By comparing how many points a team's offense earned on extra points (*OFFXP*) to the number of attempts, equal to the number of touchdowns scored by the offense, we can measure a team's extra point conversion rate (*XPCON*). In 2004 the Philadelphia Eagles scored 42 offensive touchdowns. We estimate the Eagles scored approximately 39 extra points from its offense. Hence, the Eagles *XPCON* was 93%.

With each independent variable noted, we can now fully specify our model of offensive points. This is reported in equation (14.2).

$$OFFPTS = a_{ik} + a_1 DKO + a_2 DPUNTS + a_3 DFGMISS + a_4 DINT$$
$$+ a_5 DFUMLST + a_6 START + a_7 OFFYDS + a_8 PENYDS$$
$$+ a_9 DPENYDS + a_{10} PLAYS + a_{11} 3RDCON + a_{12} FGMISS +$$
$$+ a_{13} INT + a_{14} FUMLST + a_{15} TDRATE + a_{16} XPCON. \qquad (14.2)$$

The expected signs are below each variable. Equation (14.2) was estimated with NFL team data that begins with the 1998 campaign and ends with the 2005 season. One should note that team fixed effects and dummy variables for each season were employed. The results are reported in Table 14.2.

The estimation of this model across this data reveals 91% of the variation in points scored is explained by the list of explanatory variables and furthermore, nearly all of our coefficients are of the expected sign and are statistically different from zero. The lone exceptions are *DFGMISS*, *PENYDS*, and *XPCON*. Although these variables have the expected sign, the *t*-statistics suggest none of the corresponding coefficients are statistically significant.

Fortunately, none of the insignificant factors are relevant to the story to be told about quarterbacks and running backs. Of the variables listed in equation (14.2), only four help us to evaluate the contribution of an NFL back: offensive yards, plays, interceptions thrown, and fumbles lost. Remember these four factors incorporate the impact of passing yards, rushing yards, yardage lost from sacks, passing attempts, rushing attempts, number of sacks, as well as all turnovers. All that is missing is touchdowns, a variable we cannot include since one cannot regress one variable (points) on another variable that is essentially the same (touchdowns).[§] Still, we have a measure of virtually every action a quarterback or running back takes on the field of play.

If this was baseball, one could follow the lead of Blass (1992) and stop with the modeling of offensive scoring. Unfortunately, although a quarterback only plays offense, what he does can also impact defense. Specifically, when a quarterback throws an interception or loses a fumble, he often gives the opponent advantageous field position and places additional pressure on his defense. To understand a quarterback's

---

[§]The omission of touchdown passes thrown is quite similar to measures of performance in baseball, like batting average, slugging percentage, and on-base percentage that omit runs scored and runs-batted-in.

TABLE 14.2
Modeling offensive scoring, dependent variable: Offensive Points Scored
(*OFFPTS*), team fixed effects and dummy variables for each season were
employed.

Variable	Label	Coefficient	*t*-statistic
Opponent's Kick-offs	*DKO**	0.909	3.576
Opponent's Punts	*DPUNTS***	0.448	2.206
Opponent's Missed Field Goals	*DFGMISS*	0.465	0.802
Opponent's Interceptions Thrown	*DINT**	1.272	4.339
Opponent's Fumbles Lost	*DFUMLST***	1.033	2.574
Average Starting Position of Drives	*START**	10.069	11.158
Yards Gained, Offense	*OFFYARDS**	0.080	18.688
Penalty Yards	*PENYDS*	−0.015	−1.266
Opponent's Penalty Yards	*DPENYDS**	0.055	4.927
Plays	*PLAYS**	−0.214	−4.176
Third Down Conversion Rate	*3RDCON**	1.927	3.992
Field Goals Missed	*FGMISS**	−2.986	−5.358
Interceptions Thrown	*INT**	−1.337	−3.661
Fumbles Lost	*FUMLST**	−1.481	−3.539
Percentage of Scores that are Touchdowns	*TDRATE**	102.831	4.559
Extra Point Conversion Rate	*XPCON*	45.626	1.642
Adjusted R-squared	0.91		
Observations	251		

Note: The data utilized to estimate this model came from various issues of the Official National Football League Record & Fact Book. The lone exception is *START*, which was taken from FootballOutsiders.com.
*- denotes significance at the 1% level, **- denotes significance at the 5% level.

contribution to team success we must understand how his actions impact the performance of his defense.

Fortunately, having developed a model of offense, a model of defense is essentially the same. Consider equation (14.3):

$$DEFPTS = b_{ik} + b_1KO + b_2PUNTS + b_3FGMISS + b_4INT$$
$$\qquad\qquad + \qquad\quad + \qquad\qquad + \qquad\; +$$
$$+b_5FUMLST + b_6DSTART + b_7DEFYDS + b_8DPENYDS$$
$$\quad + \qquad\qquad - \qquad\qquad + \qquad\qquad +$$
$$+b_9PENYDS + b_{10}DPLAYS + b_{11}D3RDCON + b_{12}DFGMISS +$$
$$\quad - \qquad\qquad - \qquad\qquad + \qquad\qquad -$$
$$+b_{13}DINT + b_{14}DFUMLST + b_{15}DTDRATE + b_{16}DXPCON. \qquad (14.3)$$
$$\quad - \qquad\qquad - \qquad\qquad + \qquad\qquad +$$

(The expected signs are below each variable.) Equation (14.3) includes the same collection of independent variables as our model of *OFFPTS*, except now we are focusing on the determinants of how many points a team's defense allows (*DEFPTS*). We should note that our dependent variable is a measure of how many points a defense allows. Again, points scored by the opponent via returns are not considered.

The list of independent variables begins in a familiar place, with the factors associated with the possession of the ball. When the opponent receives a kick-off (*KO*), a punt (*PUNTS*), takes possession after a missed field goal (*FGMISS*), intercepts a pass (*INT*), or recovers a fumble (*FUMLST*), it has the opportunity to score. Where the opponent begins its drive (*DSTART*) is then noted. From this starting point, an opponent moves across the field, a factor measured by its yards gained (*DYARDS*). The opponent's progress can be hampered or assisted via penalties it has incurred (*DPENYDS*) or flags thrown on your team (*PENYDS*). Again, yardage must be gained quickly before four downs have expired, so plays (*DPLAYS*) and third down conversion rates are important (*D3RDCON*). The opponent can also surrender the ball via a missed field goal (*DFGMISS*), an interception thrown (*DINT*), or losing a fumble (*DFUMLOST*). Finally, scoring is impacted by both the number of scores that are touchdowns (*DTDRATE*) and the ability to convert extra-point opportunities into points (*DXPCON*).

All of this should look quite familiar. Again, all that has been done is to take the model of *OFFPTS* and where we see a team's statistics we simply substitute the opponent's accumulation of that factor, and when we see a statistic for the opponent we use the team's variable. Given the similarities in the two models, the estimation of equation (14.3), reported in Table 14.3, holds few surprises.

Again, team fixed effects and dummy variables for each season were employed. The estimated model explains 88% of the variation in *DEFPTS* and once again virtually every coefficient is statistically different from zero. The exceptions are *FGMISS*, *DPENYDS*, and *DXPCON*. As one recalls, the corresponding factor in the model of *OFFPTS* was also statistically insignificant[¶].

[¶] Functional form might be a concern with respect to both the model of *OFFPTS* and *DEFPTS*. Specifically, should these models be estimated with a linear or double-logged functional form? This issue was examined via the Box-Cox test and the test developed by MacKinnon, White, and Davidson (1983). The

TABLE 14.3
Modeling opponent's offensive scoring, dependent variable: Opponent's Offensive Points Scored (*DEFPTS*), team fixed effects and dummy variables for each season were employed.

Variable	Label	Coefficient	t-statistic
Kick-offs	KO*	0.923	3.46
Punts	PUNTS*	0.764	3.42
Missed Field Goals	FGMISS	1.008	1.71
Interceptions Thrown	INT*	1.408	4.75
Fumbles Lost	FUMLST*	1.418	3.34
Opponent's Average Starting Position of Drives	DSTART*	9.094	9.94
Opponent's Yards Gained, Offense	DOFFYARDS*	0.078	16.38
Opponent's Penalty Yards	DPENYDS	−0.022	−1.70
Penalty Yards	PENYDS*	0.043	3.93
Opponent's Plays	DPLAYS*	−0.143	−2.76
Opponent's Third Down Conversion Rate	D3RDCON*	1.848	3.47
Opponent's Field Goals Missed	DFGMISS*	−3.425	−6.30
Opponent's Interceptions Thrown	DINT*	−1.559	−6.25
Opponent's Fumbles Lost	DFUMLST*	−1.630	−4.53
Opponent's Percentage of Scores that are Touchdowns	DTDRATE*	122.152	5.93
Opponent's Extra Point Conversion Rate	DXPCON	45.404	1.96
Adjusted R-squared	0.88		
Observations	251		

Note: The data utilized to estimate this model came from various issues of the Official National Football League Record & Fact Book. The lone exception is *DSTART*, which was taken from Football Outsiders.com.
*- denotes significance at the 1% level, **- denotes significance at the 5% level.

TABLE 14.4
Marginal value of various quarterback and running back statistics.

Variable	Impact on point differential of a one unit increase
Yards	0.080
Plays	−0.214
Interceptions	−2.745
Fumbles Lost	−2.899

14.5   NET POINTS, *QB Score*, AND *RB Score*

With models of *OFFPTS* and *DEFPTS* estimated, we can now turn to the evaluation of quarterbacks and running backs.

Let's begin with the impact of turnovers. The estimated coefficients indicate that *DEFPTS* increases by 1.559 points each time a quarterback throws an interception. From Table 14.2 we see that an interception costs the team 1.337 *OFFPTS*. Hence, an interception reduces a team's point differential by 2.745 points. Losing a fumble has a similar impact. The team loses 1.481 points while the opponent gains 1.630 points. The net impact on point differential is 2.899 points.

Contrast the impact of turnovers to the value of yards and plays. Yardage can be gained via the run or the pass. Each yard gained is worth on average 0.08 points. In contrast, a play, which can be a rushing attempt, a passing attempt, or a sack, costs the team an average of 0.214 points. For example, if a quarterback drops back to pass and is sacked for a ten yard loss, the team loses on average 0.8 points (10 × 0.08) from the lost yardage, and an additional 0.214 points on average from the sack. So the net cost is 1.01 points on average. If a quarterback then throws a twenty yard completion, the team gains an average of 1.37 points.

Given the value of yards and plays, how many yards does a quarterback have to gain to offset the impact of one interception? To gain yards the quarterback has to run a play, so we need to start with the cost of the interception and the play designed to recoup this loss. Given the value of an interception and a play, the quarterback needs to gain 2.959 points. If we divide this number by the value of a yard, we see that a quarterback needs to throw about a 37 yard pass to cancel the impact of one interception. Of course 37 yard plays are relatively rare. If it takes a quarterback five plays to overcome the cost of the interception, the quarterback will now need to gain about 50 yards to offset the cost of the interceptions and the additional plays employed.

To summarize the lessons learned from our efforts, consider Table 14.4. Here is reported the value of yards, plays, interceptions, and fumbles lost in terms of point differential, or more simply stated, net points. These values can be utilized to deter-

---

Box-Cox test indicated that the linear functional form was preferred to a double-logged form for both models. The MWD test was inconclusive with respect to the *OFFPTS* model. For the *DEFPTS* model, at the 5% level of significance, the model failed to reject the linear model. Given these tests, the linear functional form is employed.

mine how many net points are created by the numbers tabulated for quarterbacks and running backs.

Following the lead of Berri et al. (2006), though, one can build an even simpler model by noting the value of plays and turnovers in terms of yards. For example, if each yard creates 0.08 points on average, and each play costs on average 0.214 points, then each play is the equivalent of 2.69 yards. To simplify further, this ratio can be rounded to three. Similarly, the ratio of interceptions — and fumbles to yards — is 34.5 and 36.4 respectively. One can round each of these values to 35, but to make the simple model even easier to remember, it is proposed that turnovers should be valued at 30. Using these values for plays and turnovers, the following simple models — entitled *QB Score* and *RB Score* — are offered for quarterbacks and running backs.

$$QB \ Score = Yards - 3 \times Plays - 30 \times Turnovers$$
$$RB \ Score = Yards - 3 \times Plays - 30 \times FumblesLost.$$

It is important to emphasize that *QB Score* and *RB Score* are designed to be as simple as possible while still offering an accurate depiction of player performance.[∥] One might wonder if accuracy is sacrificed by using 30, as opposed to 35, for the value of turnovers. In Berri et al. (2006) the model of *OFFPTS* and *DEFPTS* did not include *START* or *DSTART*. Although the results were quite similar, the ratio of turnovers to yards was close to 50. When *START* and *DSTART* are included the ratio falls to 35. And now it is proposed that one could actually use 30 as the value of a turnover. Does it make a difference in the evaluation of quarterbacks if you move from 50, to 35, to 30?

To answer this question, every quarterback who attempted at least 100 passes was evaluated from 1995 to 2005 with three versions of *QB Score*, each utilizing a different value for turnovers. Surprisingly, the value of turnovers does not actually make any difference. The correlation coefficient between the different metrics is 0.98 and 0.99. In other words, the evaluation for quarterbacks is quite similar regardless of turnover value chosen. Consequently, it makes sense to use the value of 30, which is easier to remember given that the value associated with plays is 3.

Comparing *QB Score* to the NFL's quarterback rating system reveals some clear advantage for the former. *QB Score* is clearly an easier metric to learn and calculate. Furthermore, it incorporates what a quarterback does with his arms and legs, while the NFL's metric only considers passing statistics. Finally, the factors employed in *QB Score* are weighted in terms of the appropriate outcome, a step that is not apparently taken with the NFL's measure.

14.6   WHO IS THE BEST?

We can see how *QB Score* differs from the NFL's metric. How different are the rankings of players with each measure?

[∥] For the statistically savvy *QB Score* and *RB Score* will never be preferred to Net Points. The purpose of *QB Score* and *RB Score* is to offer a measure that is easy to calculate for people not comfortable with complicated formulas (i.e., formulas with decimals).

In Table 14.5 the top five quarterbacks in each year from 2000 to 2005 are reported. Because the NFL's metric is ultimately an efficiency measure, the quarterbacks are ranked according to *QB Score* per play, or *QB Score* divided by plays. Additionally, the ranking of these players in terms the NFL metric is also reported. Of the thirty players listed, 23 are ranked in the top five in the NFL's quarterback rating system. Of course, seven quarterbacks were not in the top five. The notable differences are Brett Favre, who ranked 10th according to the NFL's measure in 2004, but 5th in terms of *QB Score* per play. Kerry Collins in 2002 was ranked 4th in terms of *QB Score* per play but 14th of 33 signal callers according to the quarterback rating.

The other story from Table 14.5 is how different the rankings are each season. Only Peyton Manning and Trent Green appear in the table more than twice. In 2001 and 2004 three signal callers repeated top five finishes in *QB Score* per play from the previous season. But in 2003 and 2005 there were only one and two repeats respectively. Finally, no quarterback repeated from 2001 to 2002. In sum, there appears to be a fair amount of inconsistency in the performances of NFL quarterbacks.

More on inconsistency will be offered below. Before we get to that story, though, let's look at the running backs. Running backs are not evaluated in terms of a metric like the quarterback rating. Typically rankings of rushers focus strictly on rushing yards. Although such a measure is simple, it does incorporate both efficiency and durability.

Still, rushing yards alone do not incorporate receiving yards or fumbles. Hence, *RB Score* might provide a more complete picture of a running back's performance. To see this, consider Table 14.6 where the top five running backs in *RB Score* are reported from each season from 2000 to 2005.

Of the thirty players listed, ten players were not in the top five in rushing yards. The striking differences are Brian Westbrook in 2004, Charlie Garner 2002, and Tiki Barber in 2000. All were ranked below 20 in rushing yards but in the top five in *RB Score*. The difference in each case was the ability to catch the ball out of the backfield.

Beyond difference in rankings, we again see inconsistency. Only eight running backs managed to repeat top five rankings. Only Priest Holmes and LaDainian Tomlinson managed to repeat top five rankings twice. So like signal callers, we have evidence of inconsistency among the other players in the backfield.

14.7 FORECASTING PERFORMANCE IN THE NFL

As argued previously, the purpose of statistics is to separate a player from his teammates. And this is also the idea behind measuring a worker's productivity. Utilizing the data generated by the NFL and simple regression analysis we now have measures of both a quarterback's and running back's productivity. Are these measures of marginal physical product?

We can answer this question by considering an additional purpose behind collecting player statistics, i.e., the forecasting of player productivity. Decision-makers are interested in knowing which player performed better in the current campaign. Ultimately, though, decision-makers need to be able to predict who will be better next

TABLE 14.5
The top quarterbacks: 2000-2005, ranked by *QB Score* per play and the NFL's quarterback rating, minimum 224 pass attempts per season.

Quarterback	Rank *QB Score* per play	Rank *QB Rating*	Year	Yards	Plays	Turnovers	*QB Score* per play	*QB Score*	*QB Rating*
Peyton Manning	1	1	2005	3,711	503	12	1,842	3.66	104.1
Ben Roethlisberger	2	3	2005	2,325	322	10	1,059	3.29	98.6
Matt Hasselbeck	3	4	2005	3,429	509	11	1,572	3.09	98.2
Trent Green	4	8	2005	3,892	574	14	1,750	3.05	90.1
Tom Brady	5	6	2005	4,011	583	17	1,752	3.01	92.3
Peyton Manning	1	1	2004	4,494	535	11	2,559	4.78	121.1
Daunte Culpepper	2	2	2004	4,885	682	15	2,389	3.50	110.9
Donovan McNabb	3	4	2004	3,903	542	14	1,857	3.43	104.7
Trent Green	4	7	2004	4,449	613	21	1,980	3.23	95.2
Brett Favre	5	10	2004	4,023	568	18	1,779	3.13	92.4
Trent Green	1	4	2003	3,992	569	13	1,895	3.33	92.6
Peyton Manning	2	2	2003	4,186	612	11	2,020	3.30	99.0
Steve McNair	3	1	2003	3,245	457	13	1,484	3.25	100.4
Jake Plummer	4	5	2003	2,314	353	9	985	2.79	91.2
Daunte Culpepper	5	3	2003	3,705	564	17	1,503	2.66	96.4
Trent Green	1	4	2002	3,774	527	13	1,803	3.42	92.6
Chad Pennington	2	1	2002	3,034	450	7	1,474	3.28	104.2
Rich Gannon	3	2	2002	4,631	704	13	2,129	3.02	97.3
Kerry Collins	4	14	2002	3,918	613	15	1,629	2.66	85.4
Brad Johnson	5	3	2002	2,958	485	8	1,263	2.60	92.9
Kurt Warner	1	1	2001	4,657	612	26	2,041	3.33	101.4
Steve McNair	2	5	2001	3,513	543	15	1,434	2.64	90.2
Brett Favre	3	4	2001	3,826	570	21	1,486	2.61	94.1
Jeff Garcia	4	3	2001	3,678	602	15	1,422	2.36	94.8
Peyton Manning	5	8	2001	4,056	611	26	1,443	2.36	84.1
Kurt Warner	1	3	2000	3,331	385	19	1,606	4.17	98.3
Jeff Garcia	2	5	2000	4,537	657	11	2,236	3.40	97.6
Brian Griese	3	1	2000	2,651	382	7	1,295	3.39	102.9
Trent Green	4	2	2000	1,987	284	7	925	3.26	101.8
Peyton Manning	5	6	2000	4,398	628	17	2,004	3.19	94.7

TABLE 14.6
The top running backs: 2000-2005, ranked by *RB Score* per play and rushing yards.

Quarterback	Rank RB Score	Rank Rushing Yards	Year	Rushing Yards	Receiving Yards	Plays	Fumbles lost	RB Score
Tiki Barber	1	2	2005	1,860	530	411	1	1,127
Larry Johnson	2	3	2005	1,750	343	369	4	866
Shaun Alexander	3	1	2005	1,880	78	385	1	773
Warrick Dunn	4	8	2005	1,416	220	309	1	679
LaDainian Tomlinson	5	6	2005	1,462	370	390	1	632
Tiki Barber	1	5	2004	1,518	578	374	2	914
Edgerrin James	2	4	2004	1,548	483	385	2	816
Brian Westbrook	3	28	2004	812	703	250	1	735
Curtis Martin	4	1	2004	1,697	245	412	0	706
Shaun Alexander	5	2	2004	1,696	170	376	3	648
LaDainian Tomlinson	1	3	2003	1,645	725	413	0	1,131
Priest Holmes	2	9	2003	1,420	690	394	1	898
Clinton Portis	3	5	2003	1,591	314	328	1	891
Ahman Green	4	2	2003	1,883	367	405	5	885
Jamal Lewis	5	1	2003	2,066	205	413	6	852
Priest Holmes	1	3	2002	1,615	672	383	1	1,108
Charlie Garner	2	22	2002	962	941	273	0	1,084
Clinton Portis	3	4	2002	1,508	364	306	3	864
LaDainian Tomlinson	4	2	2002	1,683	489	451	1	789
Ricky Williams	5	1	2002	1,853	363	430	5	776
Marshall Faulk	1	5	2001	1,382	765	343	3	1,028
Priest Holmes	2	1	2001	1,555	614	389	3	912
Ahman Green	3	4	2001	1,387	594	366	4	763
Tiki Barber	4	19	2001	865	577	238	1	698
Garrison Hearst	5	10	2001	1,206	347	293	1	644
Marshall Faulk	1	8	2000	1,359	830	334	0	1,187
Robert Smith	2	2	2000	1,521	348	331	1	846
Edgerrin James	3	1	2000	1,709	594	450	5	803
Tiki Barber	4	22	2000	1,006	719	283	3	786
Ricky Watters	5	11	2000	1,242	613	341	2	772

TABLE 14.7

Percentage of current performance explained by past performance:
the quarterbacks.

Variable	Explanatory power
Net Points per Play	15.9%
QB Score per Play	16.7%
Completion Percentage	29.3%
Passing Yards per Pass Attempt	21.5%
Touchdowns Thrown per Pass Attempt	10.7%
Interceptions per Attempt	1.3%
Fumbles Lost per Play	5.8%
Rushing Yards per Rushing Attempt	35.6%
Rushing Touchdowns per Rushing Attempt	8.5%
Quarterback Rating	14.0%

Sample: 111 quarterbacks who attempted 224 passes in successive seasons, from 2001 to 2005. Data Source: Yahoo.com.

season. Consequently, the relationship between performance this season and last season is important.

To understand this relationship, we begin with a sample of quarterbacks who attempted at least 224 passes — the minimum needed to qualify for the NFL's quarterback rating rankings — in successive seasons from 2001 to 2005. In all, 111 quarterbacks met these criteria. With sample hand, each quarterback's performance with respect to a variety of metrics was regressed on what the player did the previous campaign.

To control for changes in games played, all statistics were considered on a per play or per attempt basis. For example, a quarterback's completion percentage — or completions per pass attempt — was regressed on his completion percentage the previous season. In all ten statistics — Net Points per Play, *QB Score* per Play, Completion Percentage, Passing Yards per Pass Attempt, Touchdowns Thrown per Pass Attempt, Interceptions per Pass Attempt, Fumbles Lost per Play, Rushing Yards per Rushing Attempt, Rushing Touchdowns per Rushing Attempt, and Quarterback Rating — were examined. The results are reported in Table 14.7.

Consistent with the results reported in Berri et al. (2006), quarterbacks appear to be quite inconsistent. Whether one looks at the individual stats — like Passing Yards per Pass Attempt or Completion Percentage — or the summary statistics like *QBScore* per Play, Net Points per Play, or the NFL's Quarterback Rating - the story is the same. The relationship between current and past performance is quite weak.**

** Weak or strong are of course relative terms. Certainly these correlations indicate that forecasting current performance with past performance would be subject to a larger forecast error. To further highlight the weakness of these correlations, though, a discussion of baseball and basketball is offered below.

TABLE 14.8

Percentage of current performance explained by past performance: the running backs.

Variable	Explanatory power
Net Points per Play	17.3%
RB Score per Play	18.5%
Rushing Yards per Rushing Attempt	9.2%
Receiving Yards per Reception	3.0%
Fumbles Lost per Play	0.6%

Sample: 95 running backs with 160 rushing attempts in successive seasons, from 2001 to 2005. Data Source: Yahoo.com.

Much of this is related to the propensity of quarterbacks to commit turnovers. The data suggests that how many interceptions a quarterback throws changes dramatically from season to season. The same story is told with respect to fumbles.

When we look at running backs, the story stays the same. These players were examined with respect to five statistics — Net Points per Play, *RB Score* per Play, Rushing Yards per Rushing Attempt, Receiving Yards per Reception, and Fumbles Lost per Play. Again, as reported in Table 14.8, we see very little consistency from year to year.

Contrast these findings in relation to what Berri et al. (2006) report for baseball and basketball. When we considered Net Points per Play for quarterbacks and running backs we see that less than 20% of current performance is explained by what the same player did last year. In baseball we can explain more than 30% of a player's current OPS — On-Base Average Plus Slugging Average — by what he did last year. For basketball we can explain 67% of a player's current Win Score — a summary measure explained in Berri et al. (2006) — by what he offered the previous season.

Of course one cannot compare directly the explanatory power of models with different dependent variables. To overcome this issue Berri et al. (2006) simply looked at how rankings of players changed from season to season. Again the story stayed the same. Quarterbacks were far more volatile in season-to-season performance relative to baseball or basketball players.

In Berri et al. (2006) it was argued that the inconsistency we see in football is tied to the interdependence we see between the performances of each player. A quarterback does not make any play by himself. Every pass is between him and a receiver. Even when he runs he requires an offensive line to block.

In contrast, baseball players bat by themselves. Even on the court we can observe basketball players doing much on their own. But football players can never make any plays by themselves. Consequently, we should not be surprised to see both quarterbacks and running backs perform inconsistently from season to season.

14.8   DO DIFFERENT METRICS TELL A DIFFERENT STORY?

Before moving on the implication of this story, let's make sure the results are not driven by our choice of metrics. The data utilized to construct Net Points, QB Score, and RB Score were the final statistics for each team in the NFL. Football Outsiders (www.footballoutsiders.com) has employed a larger data set, specifically every single play in each NFL season. In calculating Defense-adjusted Value Over Average (*DVOA*), Football Outsiders "breaks down every single play of the NFL season to see how much success offensive players achieved in each specific situation compared to the league average in that situation, adjusted for the strength of the opponent."

As the authors of this website argue, one can achieve greater significance when one considers more events. By moving from aggregate season data to play-by-play data, they argue a better picture of each player's productivity can be built.

Football Outsiders actually offers four statistics. Value Over Average (*VOA*) is like *DVOA*, except it does not adjust for strength of opponent. Like Net Points per Play, both *VOA* and *DVOA* are efficiency measures. To give credit to players who manage to stay on the field, Football Outsiders also offers Points Above Replacement (*PAR*). Again, Football Outsiders also adjusts PAR for strength of opponent, creating a metric called *DPAR*. Although the details of these metrics are beyond the scope of this paper, in essence these measures are an attempt to evaluate a player's aggregate performance on the field, as opposed to what he does per play.

To compute *VOA*, *DVOA*, *PAR*, and *DPAR* takes more effort than one has to expend to measure *QB Score* or *RB Score*. Does all this effort change the picture one paints?

Football Outsiders player evaluations cover the 2000 through 2005 seasons. If we look at all quarterbacks in this time frame who attempted at least 224 passes in a season — a sample of 195 signal callers — we see a 0.94 correlation between the rankings we derive from *VOA* and Net Points per Play. Net Points and *PAR* have a 0.93 correlation. And similar results are seen when we compare *VOA* and *PAR* to *QBScore*.

When we move to *DVOA* and *DPAR* the correlation is lowered to only 0.90. Still, there is a fairly high correlation between the metrics for quarterbacks offered by Football Outsiders and those offered for signal callers here.

When we move to running backs we see much weaker correlations. *VOA* and Net Points per play for running backs only exhibit a correlation of 0.77. For Net Points and PAR the correlation is 0.80. When we consider the metrics that account for strength of opponent, *DVOA* and *DPAR*, the correlations are even lower. Net Points and *DPAR* only have a correlation of 0.77. For Net Points per play and *DVOA* the correlation drops to 0.73. The story with respect to *RB Score* and the Football Outsiders measure is quite similar. Given these results it is clear that the Football Outsiders measures for running backs are telling a different story from what we see in Net Points and *RB Score*.

When we move on to the story of consistency, though, the story remains the same. First, consider the quarterbacks. As noted in Table 14.9, consistent with what we saw in the analysis of Net Points, *QB Score*, and the NFL's Quarterback Rating;

TABLE 14.9

Percentage of current performance explained by past performance: the football outsiders metrics.

Football Outsiders metrics	Explanatory Power for quarterbacks	Explanatory Power for running backs
VOA	11.5%	11.8%
PAR	16.5%	10.2%
DVOA	12.4%	12.1%
DPAR	19.3%	11.1%

Sample: See Tables 14.7 and 14.8. Data Source: FootballOutsiders.com.

the Football Outsiders metrics for quarterbacks indicate that performance for signal callers is quite inconsistent. And when we turn to running backs we see even less consistency. So although the Football Outsides metrics might be based on a larger data set, the picture painted does not appear to solve the problem of performance interdependency.

14.9 DO WE HAVE MARGINAL PHYSICAL PRODUCT IN THE NFL?

We have seen that applying the methods of Scully and Blass does allow one to measure performance in the NFL. But are any of the measures introduced or reviewed true measures of marginal product?

Given that these measures of performance show little consistency across time, one suspects that we do not have a metric that captures individual ability. A quarterback or running back's numbers appear to reflect his ability, as well as the ability of all those around him in the game. Consequently, we cannot argue that a player's Net Points, *QB Score*, *RB Score*, *DVOA*, *VOA*, *DPAR*, or *PAR* are suitable measures of marginal product.

This point is echoed by Fizel (2005).

> "Individual player performance is straightforward to measure in individual sports such as tennis and golf, and in a team sports such as baseball where the major confrontation is between an individual pitcher and individual batter. Individual performance metrics have also worked well in analyzing player output in the National Basketball Association. However, an individual player's performance in the NFL is highly dependent on the performance of teammates." (p. 170)

Fizel goes on to add that because of this issue, "studies of compensation, production, costs, managerial efficiency, and age-performance profiles are rare for this sport." (p. 171).

In this study the methodology offered by Scully and Blass for baseball has been applied to football. Although performance can be measured, the inconsistency we see in player performance confirms the supposition of Fizel. The evidence presented suggests that the statistics tracked for players in the NFL are not suitable for the measurement of an individual player's marginal product.

REFERENCES

Berri, D.J. (2004). A simple measure of worker productivity in the National Basketball Association. Mimeo.

Berri, D.J., S. Brook, A. Fenn, B. Frick, and R. Vicente-Mayoral (2005). The short supply of tall people: Explaining competitive imbalance in the National Basketball Association. *Journal of Economic Issues 39*(4), 1029–1041.

Berri, D.J. and E. Eschker (2005). Performance when it counts? The myth of the prime-time performer in the NBA. *Journal of Economics Issues 39*(3), 798–807.

Berri, D.J. and A. Fenn (2004). Is the sports media color-blind? Presented at the Southern Economic Association; New Orleans, Louisiana.

Berri, D.J. and A. Krautmann (2006). Shirking on the court: Testing for the disincentive effects of guaranteed pay. *Economic Inquiry 44*(3), 536–546.

Berri, D.J. and M.B. Schmidt (2006). On the road with the National Basketball Association's superstar externality. *Journal of Sports Economics 7*, 347–358.

Berri, D.J., M. Schmidt, and S. Brook (2006). *The Wages of Wins*. Stanford: Stanford University Press.

Blass, A.A. (1992). Does the baseball labor market contradict the human capital model of investment? *The Review of Economics and Statistics 74*(2), 261–268.

Fizel, J. (2005). The National Football League conundrum. In J. Fizel (Ed.), *The Handbook of Sports Economics Research*, pp. 170–171. M.E. Sharpe, Inc.

Lee, Y.H. and D.J. Berri (in press). A re-examination of production functions and efficiency estimates for the National Basketball Association. *Scottish Journal of Political Economy*.

MacKinnon, J., H. White, and R. Davidson (1983). Tests for model specification in the presence of alternative hypothesis: Some further results. *Journal of Econometrics 21*, 53–70.

NFL Record & Fact Book (Various editions). `http://www.nfl.com/history/randf`.

Scully, G.W. (1974). Pay and performance in major league baseball. *American Economic Review 64*, 915–930.

Thorn, J. and P. Palmer (1984). *The Hidden Game of Baseball: A Revolutionary Approach to Baseball and Its Statistics*. New York: Doubleday.

15

Optimal drafting in hockey pools

Amy E. Summers

Syreon Corporation

Tim B. Swartz

Simon Fraser University

Richard A. Lockhart

Simon Fraser University

ABSTRACT

This paper considers the selection of players in a common hockey pool. The selection strategy is based on the statistical modelling of the points scored by hockey players and the number of games played. A key feature of the approach is the introduction of an optimality criterion which allows drafting to be done in real time. A simulation study suggests that the approach is superior to a number of ad-hoc strategies.

15.1   INTRODUCTION

In Canada and in some of the northern states of the USA, there is a general excitement each spring, not only with the change in weather but also with the advent of the Stanley Cup Playoffs. A popular activity among friends and co-workers is the participation in a Stanley Cup Playoff pool.

A common Stanley Cup Playoff pool proceeds along the following lines. Among K drafters (participants in the pool), an order is determined from which drafter 1 selects a hockey player from any of the 16 teams in the National Hockey League (NHL) that have qualified for the Stanley Cup Playoffs. Drafter 2 then selects a player but is not allowed to select the player chosen by drafter 1. The drafting continues until the first round is complete (i.e., drafter K has made a selection). The order of drafting is then reversed for the second round, and the process continues for m rounds. At the completion of the draft, each drafter has selected a *lineup* of m players where each player accumulates points (i.e., goals plus assists) during the playoffs. The drafter whose lineup has the greatest number of total points is declared the winner. Typically, a monetary prize is given to the winner.

Now, the question arises as to how one should select hockey players. Clearly, players who are able to generate lots of points in a game have some appeal. However, this must be tempered by the strength of a player's team. For example, a weak team is likely to be eliminated early in the Stanley Cup Playoffs, and therefore, a good player on a weak team may not be as valuable as a weaker player on a stronger team. One might also consider the effect of the "eggs in one basket" syndrome. By choosing players predominantly from one team, a drafter's success is greatly influenced by the success of the team. It is fair to say that it is not obvious how to best select hockey players in a draft.

Although we have not come across any previous work concerning drafting in hockey pools, there is a considerable literature on the related problems of rating sports teams and predicting the outcome of sporting events. For example, Berry, Reese, and Larkey (1999) compare players of different eras in the sports of professional hockey, golf, and baseball. Carlin (1996) uses point spreads to estimate prediction probabilities for the NCAA basketball tournament. More generally, the volume edited by Bennett (1998) covers a wide range of topics related to statistical issues in sport.

This paper considers a statistical approach to the player selection problem in playoff hockey pools. More detail on all aspects of the proposed approach can be found in Summers (2005). In section 15.2, some statistical modelling is proposed for the number of points scored and the number of games played by hockey players. Together with Sportsbook odds, subjective probabilities, connected graphs, Newton-Raphson optimization, and simulation, expectations concerning the total points by lineups are obtained. A key point is that the expectations are calculated in advance of the draft so that drafting may be done in real time. Friends and co-workers may not be entirely understanding if they need to wait long periods of time for a drafter to make a selection. In section 15.3, an optimality criterion is introduced for the selection of hockey players, and the optimality criterion is a simple function of the expectations derived in section 15.2. In section 15.4, we conduct a simulation study to assess the proposed selection strategy against some common ad-hoc strategies. We observe that the proposed selection strategy is arguably the best strategy. The results of an actual Stanley Cup playoff pool using our methodology are reported in section 15.5. We conclude with a short discussion in section 15.6.

15.2   STATISTICAL MODELLING

15.2.1   DISTRIBUTION OF POINTS

Consider a hockey player who is on the ice and whose team is in possession of the puck. We let p denote the probability that the team scores and the player accumulates a point during the possession. If there are n such possessions during the game, then it may be reasonable to model the number of points X obtained by the player during the game as a Binomial(n, p) random variable. Naturally, this is a simplification which assumes independence of possessions, does not take account of power plays, does not take account of teammates and opposition, etc. When n is large and p is small, we apply the standard approximation

$$X \sim \text{Poisson}(\theta),$$

where $\theta = np$. The Poisson distribution has been previously used for modelling goals scored in hockey games and has provided a good fit (Berry, 2000). The Poisson is preferable to the Binomial as it is characterized by a single parameter.

A straightforward method for estimating θ is

$$\hat{\theta} = \frac{\text{number of points obtained by the player in the regular season}}{\text{number of games played by the player in the regular season}}. \qquad (15.1)$$

A drafter may choose to modify the estimator (15.1) in various ways such as assigning a greater weight to recent performance or taking into account previous playoff performance. For ease of notation, all subsequent usages of θ will actually correspond to the estimated $\hat{\theta}$.

Consider now the number of points Y_{ki} obtained during the entire Stanley Cup Playoffs by the ith player selected by drafter k. If the player's θ value is θ_{ki} and he plays in g_{ki} playoff games, it follows that

$$Y_{ki} \mid g_{ki} \sim \text{Poisson}(g_{ki}\theta_{ki}).$$

We then define the total number of points obtained by lineup k

$$T_k \equiv Y_{k1} + \cdots + Y_{km},$$

and note that the unconditional distribution of T_k (i.e., unconditional with respect to $g_{k1}, \ldots, g_{km}$) is the sum of m dependent Poisson mixture distributions. Using the conditional expectation formulae and assuming that Y_{ki} and Y_{lj} are conditionally independent given g_{ki} and g_{lj} for all k, i, l, j, we obtain

$$E(T_k) = \sum_{i=1}^{m} \theta_{ki} E(g_{ki}), \qquad (15.2)$$

$$\text{var}(T_k) = \sum_{i=1}^{m} \theta_{ki} E(g_{ki}) + \sum_{i=1}^{m} \theta_{ki}^2 \text{var}(g_{ki}) +$$
$$2 \sum_{i<j} \theta_{ki} \theta_{kj} \text{cov}(g_{ki}, g_{kj}), \qquad (15.3)$$

and

$$\text{cov}(T_k, T_l) = \sum_{i=1}^{m} \sum_{j=1}^{m} \theta_{ki} \theta_{lj} \text{cov}(g_{ki}, g_{lj}). \qquad (15.4)$$

The expressions (15.2), (15.3), and (15.4) are used in section 15.3 to determine an optimal player selection strategy. The terms in the expressions are all known except for the expectations involving the number of games played, and these expectations are obtained via simulation in section 15.2.2. Admittedly, the conditional independence assumption may be somewhat questionable in the case of two players from the same line on the same team.

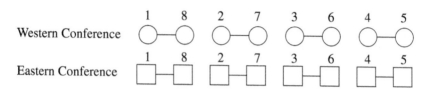

FIGURE 15.1

First round format.

15.2.2 DISTRIBUTION OF GAMES

If we are able to assign probabilities for each of the playoff teams defeating each of the other playoff teams in all potential playoff games, then it is possible to simulate the entire NHL playoffs. The simulation matches teams in the first round as dictated by the actual NHL playoff matchups, and depending on which teams win the simulated series, team matchups are determined for the second round, and so on, leading to the two Stanley Cup finalists. By repeating the playoff simulation many times, the expectations appearing in the expressions (15.2), (15.3), and (15.4) can be estimated accurately using simple averages.

We now use Sportsbook odds, subjective probabilities, connected graphs, and Newton-Raphson optimization to assign the individual game probabilities. We begin with the preliminary task of assigning the *series* probabilities $P(i, j)$, $i \neq j$ where $P(i, j)$ denotes the probability that team i defeats team j in a best of seven series. A good way to assign series probabilities is to utilize Sportsbook odds as these are known to reflect the collective wisdom of the betting public (Insley, Mok, and Swartz, 2004). Sportsbook odds are reported in the form odds$(i, j) : 1$ where odds(i, j) is the payout in dollars on a winning one dollar bet on team i and $P(i, j) = 1/(\text{odds}(i, j) + 1)$. Observe that odds$(i, j) \cdot \text{odds}(j, i) = 1$. A difficulty is that prior to the Stanley Cup Playoffs (when the draft takes place), Sportsbook odds are only available for actual first round matchups, and not all of the hypothetical matchups that may occur later in the playoffs. Figure 15.1 gives the layout for the first round of the playoffs where the numbers correspond to the seedings of the 8 teams in both the Eastern and Western conferences. Each line in the graph indicates that the probability $P(i, j)$ between the two connected teams is available via Sportsbook odds.

To complete the unspecified probabilities $P(i, j)$ where $P(j, i) = 1 - P(i, j)$, we use the drafter's subjective hockey knowledge. The dotted lines in Figure 15.2 represent two of the drafter's subjective probabilities. Fortunately, it is not necessary for the drafter to assign all of the $\binom{16}{2} - 8 = 112$ remaining series probabilities. By assuming that the probabilities $P(i, j)$ are "transitive," we can use the Sportsbook odds and the drafter's subjective probabilities to determine the probabilities of other matchups. For example, referring to Figure 2, $P(8E, 7W)$ can be obtained by following the line from $8E$ to $7W$ whereby

$$\text{odds}(8E, 7W) = \text{odds}(8E, 2W) \cdot \text{odds}(2W, 7W).$$

The goal is to create a connected graph; a graph is connected if there exists a path between each pair of vertices. In this situation, the drafter must specify a minimum of seven subjective probabilities in order to connect the graph and thereby complete all probabilities $P(i, j), i \neq j$.

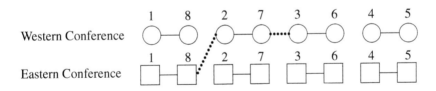

FIGURE 15.2

Drafter's probabilities dotted.

Of course, if one is going to complete the probabilities via transitivity, there should not exist different paths that lead to different probability calculations. The drafter should be "transitivity coherent" in his or her subjective probability assignments. For example, $P(1, 4) = 0.4$ and $P(4, 5) = 0.5$ imply odds$(1, 4) = 1.5$ and odds$(4, 5) = 1.0$ which is transitivity incoherent with respect to $P(1, 5) = 0.8$ (i.e., odds$(1, 5) = 0.25$). Note that transitivity is a strong assumption that is not always applicable in sports, particularly when style of play and individual player matchups are paramount. However, we believe that transitivity is fairly sensible in hockey. If a drafter is adamant that transitivity is inappropriate with respect to some teams, the drafter should simply assign the corresponding entries $P(i, j)$ rather than impute the $P(i, j)$ entries via transitivity.

Having determined the series probabilities $P(i, j)$, it is now necessary to calculate individual game probabilities. To this end, let p_{ij} denote the probability that team i defeats team j in a game on neutral ice, and let ϵ denote the home ice advantage. Therefore, with respect to team i defeating team j, $p_{ij} + \epsilon$ is the probability of a home win, and $p_{ij} - \epsilon$ is the probability of an away win. We obtain an estimate of ϵ common to the league by considering the results of the regular season where

$$\epsilon + 1/2 = \frac{\text{(number of home team wins)} + \frac{1}{2} \text{ (number of tied games)}}{\text{total number of regular season games}}. \quad (15.5)$$

Keeping in mind the 2-2-1-1-1 home/away format for NHL playoff series, let team i denote the higher seeded team for whom the first two games in the series are played at home. Then

$$P(i, j) = \text{Pr}(i \text{ wins in 4 games}) + \text{Prob}(i \text{ wins in 5 games}) +$$
$$\text{Pr}(i \text{ wins in 6 games}) + \text{Pr}(i \text{ wins in 7 games}), \quad (15.6)$$

where

$$\Pr(i \text{ wins in 4 games}) = (p_{ij} + \epsilon)^2 (p_{ij} - \epsilon)^2, \tag{15.7}$$

$$\Pr(i \text{ wins in 5 games}) = 2(p_{ij} + \epsilon)^2 (p_{ij} - \epsilon)^2 (1 - (p_{ij} + \epsilon)) +$$
$$2(p_{ij} + \epsilon)^3 (p_{ij} - \epsilon)(1 - (p_{ij} - \epsilon)), \tag{15.8}$$

$$\Pr(i \text{ wins in 6 games}) = 3(p_{ij} + \epsilon)(p_{ij} - \epsilon)^3 (1 - (p_{ij} + \epsilon))^2 +$$
$$(p_{ij} + \epsilon)^3 (p_{ij} - \epsilon)(1 - (p_{ij} - \epsilon))^2 +$$
$$6(p_{ij} + \epsilon)^2 (p_{ij} - \epsilon)^2 (1 - (p_{ij} + \epsilon)) \times$$
$$(1 - (p_{ij} - \epsilon)), \tag{15.9}$$

$$\Pr(i \text{ wins in 7 games}) = (p_{ij} + \epsilon)(p_{ij} - \epsilon)^3 (1 - (p_{ij} - \epsilon))^3 +$$
$$(p_{ij} + \epsilon)^4 (1 - (p_{ij} - \epsilon))^3 +$$
$$9(p_{ij} + \epsilon)^2 (p_{ij} - \epsilon)^2 (1 - (p_{ij} + \epsilon))^2 (1 - (p_{ij} - \epsilon)) +$$
$$9(p_{ij} + \epsilon)^3 (p_{ij} - \epsilon)(1 - (p_{ij} + \epsilon)) \times$$
$$(1 - (p_{ij} - \epsilon))^2. \tag{15.10}$$

We then substitute (15.7), (15.8), (15.9), and (15.10) into (15.6) where $P(i, j)$ and ϵ are both known. We therefore have an equation with only one unknown p_{ij}, and p_{ij} can be calculated using the Newton-Raphson algorithm. Substituting p_{ij} back into (15.7), (15.8), (15.9) and (15.10) gives us the four probabilities corresponding to team i winning the series against team j. Noting that $p_{ji} = 1 - p_{ij}$ and using formulae (15.7)–(15.10), we can directly calculate the four probabilities corresponding to team j winning the series against team i.

In summary, probability distributions with eight outcomes are obtained for all 120 possible series. Using these distributions, the playoffs are simulated in advance of the draft to estimate the expectations appearing in the expressions (15.2), (15.3), and (15.4). The expressions (15.2), (15.3), and (15.4) can then be rapidly calculated for various lineups during the draft.

15.3   AN OPTIMALITY CRITERION

Without loss of generality, suppose that drafter j uses the following player selection strategy. For the first round of the draft, we suggest that drafter j select the player among those available whose expected number of points in the playoffs $\theta E(g)$ is greatest. Recall that θ and $E(g)$ have been determined in advance of the draft for each player, and therefore, the decision can be made in real time.

In round n of the draft, $2 \leq n \leq m$, we propose that drafter j choose the player whose selection maximizes

$$LD = \sum_{i \neq j} \Pr(T_j > T_i), \tag{15.11}$$

where $T_1, \ldots, T_K$ correspond to the current (yet incomplete) lineups in round n. We refer to LD in (15.11) as the optimality criterion and interpret LD as the expected number of lineups that are defeated by lineup j. Although maximizing the function

$$\Pr\left(\bigcap_{i \neq j}(T_j > T_i)\right),$$

may also seem appealing, we have a simple approximation of LD, and we are able to maximize the approximation of LD in real time.

On the website www.statistical-thinking-in-sports.com, Appendix A of this paper is located, and there we suggest the approximation

$$T_j - T_k \sim \text{Normal}(\mu_{jk}, \sigma_{jk}^2), \tag{15.12}$$

where $\mu_{jk} = E(T_j) - E(T_k)$, $\sigma_{jk}^2 = \text{var}(T_j) + \text{var}(T_k) - 2\text{cov}(T_j, T_k)$, and we refer to expressions (15.2), (15.3), and (15.4) which are easily modified for rounds $2 \leq n \leq m$ of the draft. Therefore LD in (15.11) is approximated by

$$\sum_{j \neq i} \Phi\left(\frac{\mu_{ji}}{\sigma_{ji}}\right), \tag{15.13}$$

where Φ is the cumulative distribution function of the standard normal distribution. A key point is that (15.13) can be evaluated (by computer software) for each of the unselected hockey players in round n, and in real time, we can choose that hockey player whose selection maximizes (15.13).

A further simplification may be possible in terms of the search over the remaining pool of hockey players. If we order the players on each of the 16 playoff hockey teams in decreasing order according to their θ values, then our intuition suggests that only the highest player remaining on each of the 16 teams needs to be considered in the selection process. This reduces the search to a set of only 16 players. The idea is that we never prefer a player over a teammate if the first player's ability to score points is less than the second player's ability to score points. The conjecture is addressed in Appendix B of this paper located at the website www.statistical-thinking-in-sports.com.

15.4   A SIMULATION STUDY

We consider an NHL playoff hockey pool with $K = 10$ drafters and $m = 10$ rounds of drafting. Our pool is based on results from the 2003-2004 NHL regular season where we obtain the player characteristics θ using (15.1), and we estimated the home ice advantage $\epsilon = 0.05$ as in (15.5). For convenience, we restrict our attention to the 10 players per playoff team having the highest θ values. This provides 160 eligible players among whom 100 are selected in the draft. The series win probabilities $P(i, j)$ for all $i \neq j$ are obtained using the transitivity assumption based on the subset of Sportsbook and subjective probabilities given in Table 15.1.

Using the series win probabilities $P(i, j)$, the corresponding single game win probabilities p_{ij} are then calculated. We then simulate the playoffs to obtain the

TABLE 15.1

Sportsbook and subjective series win
probabilities $P(i, j)$ for team i defeating
team j.

Team i	Team j	$P(i, j)$
Detroit	Nashville	0.84
San Jose	St. Louis	0.58
Vancouver	Calgary	0.56
Colorado	Dallas	0.55
Tampa Bay	NY Islanders	0.69
Boston	Montreal	0.60
Philadelphia	New Jersey	0.52
Toronto	Ottawa	0.48
Detroit	Vancouver	0.61
Vancouver	Colorado	0.52
Tampa Bay	Boston	0.52
Boston	Vancouver	0.53
Calgary	Ottawa	0.43
Detroit	San Jose	0.56
Calgary	New Jersey	0.45

quantities $E(g_{ki})$, var(g_{ki}), and cov(g_{ki}, g_{lj}) appearing in (15.2), (15.3), and (15.4). These quantities relate to the number of games played by each player in the playoffs and are needed to implement our optimal drafting strategy. Table 15.2 gives the top 10 players according to their expected number of playoff points where we recall that this quantity is a product of the player's θ and $E(g)$ values. We note that the top 10 players belong to only four different teams.

To simulate a playoff pool, we need to create "virtual" drafters. The drafters adhere to the following rules in selecting their lineups.

- Drafter 1 $\sim$ chooses players according to the optimality criterion (15.13).

- Drafter 2 $\sim$ chooses players with the largest θ values.

- Drafter 3 $\sim$ chooses players with the largest expected number of playoff points.

- Drafter 4 $\sim$ is an advocate of numerology. The drafter believes that the numbers 8 and 9 are lucky and chooses players with the most regular season points that are divisible by 8 or 9.

- Drafter 5 $\sim$ roots for the underdog by choosing players with the most points alternating between the lowest seeded teams in the Eastern and Western conferences.

- Drafter 6 $\sim$ roots for the favorite by choosing players with the most points alternating between the highest seeded teams in the Eastern and Western conferences.

TABLE 15.2

Top 10 hockey players according to their expected number of playoff points.

Player	Team	θ	Expected Points
Robert Lang	Detroit	1.15	17.21
Peter Forsberg	Colorado	1.41	15.35
Martin St. Louis	Tampa Bay	1.15	14.89
Pavel Datsyuk	Detroit	0.91	13.63
Cory Stillman	Tampa Bay	0.99	12.83
Brett Hull	Detroit	0.84	12.62
Brad Richards	Tampa Bay	0.96	12.52
Alex Tanguay	Colorado	1.15	12.46
Markus Naslund	Vancouver	1.08	12.15
Joe Sakic	Colorado	1.07	11.69

- Drafter 7 ~ chooses players with the most regular season points.

- Drafter 8 ~ is a Vancouver Canucks "Superfan." The drafter chooses only Canucks and chooses Canucks according to the highest number of regular season points.

- Drafter 9 ~ chooses players with the most regular season points whose first name begins with the letter S.

- Drafter 10 ~ chooses players with the highest θ values from the top four seeded teams in the first four rounds of the draft. For the remaining six rounds, the drafter chooses players with the highest θ values.

Both drafter 1 (who uses the optimality criterion) and drafter 3 have an advantage. The advantage is that they have exact knowledge of an aspect of the simulation, namely the number of expected playoff points for each player. We suspect that it would be unlikely for an office colleague to go to the extreme lengths of implementing the selection strategy of drafter 3. Nevertheless, we want to include a drafter who is "smart" for the sake of assessing the optimality criterion. We also remark that many of the above drafters are fairly sophisticated in their selection strategies. Our experience is that drafters are generally not this sophisticated. Therefore, our proposed approach may work better in practice than in the following test simulation.

We now consider a simulation where each drafter drafts in the position according to their number. The playoffs were simulated 10000 times and the order of finish in each simulation was recorded for each of the drafters. In Table 15.3, we provide the cumulative probabilities for finishing in positions 1, ..., 10 in the playoff pool. We also provide $E(LD)$, the expected number of lineups that each drafter defeats. We observe that $E(LD) = 6.54$ for drafter 1, and this is the largest expectation among all drafters. This might be expected as the optimality criterion attempts to maximize the number of lineups defeated. However, we note that drafter 1 does not have the

TABLE 15.3

Cumulative probabilities for order of finish and $E(LD)$ in a playoff pool where drafter 1 selects first.

Finish	Drafter									
	1	2	3	4	5	6	7	8	9	10
1st	0.11	0.21	0.15	0.02	0.01	0.16	0.07	0.19	0.00	0.06
2nd	0.32	0.32	0.36	0.10	0.03	0.29	0.20	0.24	0.01	0.15
3rd	0.56	0.40	0.57	0.16	0.05	0.36	0.34	0.26	0.02	0.27
4th	0.77	0.50	0.72	0.24	0.06	0.45	0.50	0.30	0.04	0.45
5th	0.88	0.61	0.81	0.34	0.09	0.55	0.67	0.33	0.07	0.66
6th	0.94	0.75	0.88	0.50	0.11	0.65	0.81	0.39	0.13	0.84
7th	0.98	0.87	0.94	0.72	0.14	0.73	0.92	0.47	0.30	0.94
8th	0.99	0.95	0.98	0.93	0.20	0.82	0.98	0.53	0.64	0.99
9th	1.00	1.00	1.00	0.99	0.44	0.91	1.00	0.71	0.96	1.00
10th	1.00	1.00	1.00	1.00	1.00	1.00	1.00	1.00	1.00	1.00
$E(LD)$	6.54	5.60	6.40	3.99	1.12	4.92	5.49	3.41	2.17	5.35

highest probability of finishing first in the hockey pool. This honor belongs to drafter 2, followed by drafters 8, 6, and 3. In this regard, the success of drafter 8 (the Canucks Superfan) provides some evidence that "putting all of one's eggs in one basket" has some merit. When the Canucks go deep into the playoffs, drafter 8 does very well, and conversely, when the Canucks are eliminated early in the playoffs, drafter 8 does poorly. We also note that the probability of "being in the money" (i.e., first, second, or third) is the greatest for drafter 3 followed closely by drafter 1. Drafter 3 (1) is in the money 57% (56%) of the time. As many hockey pools have prizes for first, second, and third place finishes, this provides further appeal for taking into account the number of games played.

We now consider the impact of draft position. We repeat the simulation where all of the drafters maintain the same order of selection except drafter 1 who is moved from the first to the last position. This means, for example, that drafter 1 has the 10th and the 11th selections in the draft. The results of the simulation are provided in Table 15.4. We observe that drafter 1 does about as well selecting in the last position as in the first position. For example, drafter 1 finishes second to drafter 3 in $E(LD)$ and drafter 1 is second according to being in the money. Again, we are not too disappointed by the $E(LD)$ result as drafter 3 uses specialized knowledge that would not be typically available to drafters. Our explanation for this is that drafting in the 10th position is too far away from the beginning of the draft to select one of the few really outstanding hockey players.

Now recall the discussion where it was pointed out that drafter 1 and drafter 3 were advantaged by their knowledge of the expected number of playoff points. It therefore might be asked whether there is a meaningful difference in their selection strategies. In other words, does it make sense to invoke the optimality criterion when a similar lineup might be achieved by the simpler criterion of maximizing the expected number of playoff points? To investigate this, we consider whom drafter 3 would

TABLE 15.4

Cumulative probabilities for order of finish and $E(LD)$ in a playoff pool where drafter 1 selects last.

Finish	Drafter									
	1	2	3	4	5	6	7	8	9	10
1st	0.13	0.13	0.18	0.02	0.01	0.22	0.07	0.18	0.00	0.05
2nd	0.31	0.25	0.46	0.08	0.03	0.32	0.19	0.23	0.01	0.14
3rd	0.50	0.36	0.64	0.15	0.05	0.39	0.34	0.27	0.01	0.29
4th	0.67	0.51	0.75	0.24	0.06	0.46	0.49	0.30	0.03	0.48
5th	0.79	0.66	0.84	0.36	0.09	0.54	0.64	0.34	0.07	0.69
6th	0.87	0.83	0.90	0.51	0.12	0.62	0.77	0.40	0.14	0.85
7th	0.93	0.94	0.95	0.72	0.15	0.70	0.88	0.47	0.33	0.94
8th	0.97	0.98	0.98	0.93	0.20	0.80	0.97	0.53	0.65	0.99
9th	0.99	1.00	1.00	0.99	0.44	0.91	1.00	0.71	0.97	1.00
10th	1.00	1.00	1.00	1.00	1.00	1.00	1.00	1.00	1.00	1.00
$E(LD)$	6.16	5.66	6.70	3.99	1.14	4.96	5.33	3.43	2.21	5.42

have selected had drafter 3 been in the position of drafter 1. In the simulation where drafter 1 chose first, drafter 3 would have picked the same player as drafter 1 in 8 of the 10 draft positions. In the simulation where drafter 1 chose last, drafter 3 would have picked the same player as drafter 1 in only 6 of the 10 draft positions. Hence, the strategy of drafter 1 based on the optimality criterion does differ meaningfully from the simplified (but intelligent) strategy of drafter 3.

15.5 AN ACTUAL STANLEY CUP PLAYOFF POOL

Although a particular instance of a hockey pool does not address the long term properties of a drafting strategy, it is instructive to review our approach in a realistic setting. For the 2006 NHL playoffs, we conducted a pool with $K = 10$ drafters and $m = 10$ rounds of drafting. Our strategy was implemented by one of the authors (Amy), who by chance, drafted in the first position.

In Table 15.5, we provide the Sportsbook odds (adjusted for vigorish) for each of the favorites in the eight first round series. The odds were taken from www. vegasinsider.com two days prior to the beginning of the playoffs. The underdog is listed in parentheses. In Table 15.6, we provide Amy's subjective odds for seven hypothetical matchups. Some of the odds may seem extreme, but this demonstrates the flexibility of the approach. All of the odds in Table 15.5 and Table 15.6 were then used to complete the graph and populate the probability matrix.

For the sake of brevity, in Table 15.7, we show the first 3 draft picks for each of the drafters, the draft order, playoff points accumulated by the draft picks, and the total number of playoff points for the entire lineup. The 2006 playoffs were marked by upsets with Carolina defeating Edmonton in the Stanley Cup finals. However, by skill and good fortune, Amy won the hockey pool with 113 points followed by Willy with 97 points.

TABLE 15.5

Sportbook odds for the favorite in
the first round series (underdog in
parentheses).

Teams	Odds
Detroit (Edmonton)	0.37
Dallas (Colorado)	0.61
Calgary (Anaheim)	0.61
Nashville (San Jose)	0.92
Ottawa (Tampa Bay)	0.41
Carolina (Montreal)	0.56
New Jersey (New York)	0.54
Buffalo (Philadelphia)	0.67

TABLE 15.6

Subjective odds for the favorite in
hypothetical matchups (underdog
in parentheses).

Teams	Odds
Dallas (Detroit)	0.54
Calgary (Detroit)	0.43
Calgary (Nashville)	0.25
Anaheim (New York)	0.18
Anaheim (Philadelphia)	0.82
Carolina (Ottawa)	0.92
Calgary (Carolina)	0.75

TABLE 15.7
Some summary results from the 2006 NHL playoff pool.

Drafter	Pick	Points	Drafter	Pick	Points
Amy (1)	P Elias (NJ)	16	Jen (2)	J Jagr (NY)	1
	D Briere (Buf)	19		P Schaefer (Ott)	7
	C Stillman (Car)	26		J Iginla (Cgy)	8
Total Points = 113			Total Points = 51		
Darcy (3)	D Heatley (Ott)	12	Leslie (4)	E Staal (Car)	28
	P Kariya (Nsh)	7		B Gionta (NJ)	7
	M Modano (Dal)	4		R Lang (Det)	6
Total Points = 51			Total Points = 93		
Linda (5)	J Thornton (SJ)	9	AJ (6)	B Shanahan (Det)	2
	N Lidstrom (Det)	2		J Spezza (Ott)	14
	T Holmstrom (Det)	3		R Smyth (Edm)	16
Total Points = 49			Total Points = 80		
Beth (7)	A McDonald (Ana)	9	Willy (8)	H Zetterberg (Det)	6
	A Tanguay (Col)	6		P Datsyuk (Det)	3
	A Hemsky (Edm)	17		P Marleau (SJ)	14
Total Points = 57			Total Points = 97		
Art (9)	D Alfredsson (Ott)	10	Rory (10)	J Cheechoo (SJ)	9
	T Selanne (Ana)	14		M Afinogenov (Buf)	8
	S Gomez (NJ)	9		P Forsberg (Phi)	8
Total Points = 93			Total Points = 67		

15.6 DISCUSSION

We have proposed a method of player selection in a common NHL hockey pool. With the aid of a computer and some preliminary simulations, player selections can be made in real time. It appears that the approach is better than a number of ad-hoc strategies.

It is clear that the approach could also be implemented for other hockey leagues (e.g., junior hockey), and it might also be tweaked for pools with slightly different rules and also for different sports (e.g., soccer).

REFERENCES

Bennett, J. (Ed.) (1998). *Statistics in Sport.* New York: Oxford University Press.

Berry, S.M. (2000). My triple crown. *Chance 13*(2), 56–61.

Berry, S.M., C.S. Reese, and P.D. Larkey (1999). Bridging different eras in sports. *Journal of the American Statistical Association 94*(447), 661–676.

Carlin, B.P. (1996). Improved NCAA basketball tournament modeling via point spread and team strength information. *The American Statistician 50*, 39–43.

Insley, R., L. Mok, and T.B. Swartz (2004). Issues related to sports gambling. *The Australian and New Zealand Journal of Statistics 46*, 219–232.

Summers, A. (2005). Hockey pools for profit: a simulation based player selection strategy. Master's thesis, Department of Statistics and Actuarial Science, Simon Fraser University, Burnaby BC, Canada.

References

Albert, J. (2001). *Using play-by-play baseball data to develop a better measure of batting performance.* Bowling Green State University. Online at `bayes.bgsu.edu/papers/rating_paper2.pdf`.

Albert, J. (2002). Hitting with runners in scoring position. *Chance 15*, 8–16.

Albert, J. (2003). *Teaching Statistics Using Baseball.* Washington, DC: Mathematical Association of America.

Albert, J. (2005). Does a baseball hitter's batting average measure ability or luck? *Stats 44.*

Albert, J. and J. Bennett (2003). *Curve Ball: Baseball, Statistics, and the Role of Chance in the Game* (revised ed.). New York: Springer-Verlag.

Albright, S.C. (1993). A statistical analysis of hitting streaks in baseball. *Journal of the American Statistical Association 88*, 1175–1183 (with discussion).

Anderson, C.L. (1977). Note on the advantage of first serve. *Journal of Combinatorial Theory A 23*, 363.

Audas, R., S. Dobson, and J. Goddard (2002). The impact of managerial change on team performance in professional sports. *Journal of Economics and Business 54*, 633–650.

Baimbridge, M., S. Cameron, and P. Dawson (1996). Satellite television and the demand for football: A whole new ball game? *Scottish Journal of Political Economy 43*, 317–333.

Barnett, V. and S. Hilditch (1993). The effect of an articifial pitch surface on home team performance in football (soccer). *Journal of the Royal Statistical Society, A 156*, 39–50.

Bennett, J. (Ed.) (1998). *Statistics in Sport.* New York: Oxford University Press.

Berri, D.J. (2004). A simple measure of worker productivity in the National Basketball Association. Mimeo.

Berri, D.J., S. Brook, A. Fenn, B. Frick, and R. Vicente-Mayoral (2005). The short supply of tall people: Explaining competitive imbalance in the National Basketball Association. *Journal of Economic Issues 39*(4), 1029–1041.

Berri, D.J. and E. Eschker (2005). Performance when it counts? The myth of the prime-time performer in the NBA. *Journal of Economics Issues 39*(3), 798–807.

Berri, D.J. and A. Fenn (2004). Is the sports media color-blind? Presented at the Southern Economic Association; New Orleans, Louisiana.

Berri, D.J. and A. Krautmann (2006). Shirking on the court: Testing for the disincentive effects of guaranteed pay. *Economic Inquiry 44*(3), 536–546.

Berri, D.J. and M.B. Schmidt (2006). On the road with the National Basketball Association's superstar externality. *Journal of Sports Economics 7*, 347–358.

Berri, D.J., M.B. Schmidt, and S.L. Brook (2004). Stars at the gate: The impact of star power on NBA gate revenues. *Journal of Sports Economics 5*(1), 33–50.

Berri, D.J., M. Schmidt, and S. Brook (2006). *The Wages of Wins*. Stanford: Stanford University Press.

Berry, S.M. (2000). My triple crown. *Chance 13*(2), 56–61.

Berry, S.M., C.S. Reese, and P.D. Larkey (1999). Bridging different eras in sports. *Journal of the American Statistical Association 94*(447), 661–676.

von Bertalanffy, L. (1938). A quantitative theory of organic growth. *Human Biology 10*, 181–213.

Bissinger, B. (2005). *Three Nights in August: Strategy, Heartbreak, and Joy Inside the Mind of a Manager*. Mariner Books.

Blass, A.A. (1992). Does the baseball labor market contradict the human capital model of investment? *The Review of Economics and Statistics 74*(2), 261–268.

Blest, D.C. (1996). Lower bounds for athletic performances. *The Statistician 45*, 243–253.

Borghans, L. (1995). Keuzeprobleem op Centre Court. *Economisch Statistische Berichten 80*, 658–661.

Borland, J. and R. Macdonald (2003). Demand for sport. *Oxford Review of Economic Policy 19*, 478–502.

Boulier, B. and H. Stekler (2003). Predicting the outcomes of National Football League games. *International Journal of Forecasting 19*, 257–270.

Brain, P. and R. Cousens (1989). An equation to describe dose responses where there is stimulation of growth at low dose. *Weed Research 29*, 93–96.

Bray, S.R., J. Obrara, and M. Kwan (2005). Batting last as a home advantage factor in men's NCAA tournament baseball. *Journal of Sports Sciences 23*(7), 681–686.

Burnham, K.P. and D.R. Anderson (2002). *Model Selection and Multimodel Inference: A Practical Information-Theoretic Approach* (2nd. ed.). New York: Springer-Verlag.

Cain, M., D. Law, and D. Peel (2000). The favourite-longshot bias and market efficiency in UK football betting. *Scottish Journal of Political Economy 47*, 25–36.

Cairns, J. (1990). The demand for professional team sports. *British Review of Economic Issues 12*(28), 1–20.

Carlin, B.P. (1996). Improved NCAA basketball tournament modeling via point spread and team strength information. *The American Statistician 50*, 39–43.

Charnes, A., W.W. Cooper, and E. Rhodes (1978). Measuring the efficiency of decision making units. *European Journal of Operational Research 2*, 429–444.

Chatterjee, S. and S. Chatterjee (1982). New lamps for old: An exploratory analysis of running times in Olympic Games. *Applied Statistics 31*, 14–22.

Clarke, S.R. (1993). Computer forecasting of Australian Rules Football for a daily newspaper. *Journal of the Operational Research Society 44*, 753–799.

Clarke, S.R. and J.M. Norman (1995). Home ground advantage of individual clubs in English soccer. *The Statistician 44*(4), 509–521.

Courneya, K.S. and A.V. Carron (1990). Batting first versus last: Implications for the home advantage. *Journal of Sport and Exercise Psychology 12*, 312–316.

Cramer, R.D. (1977). Do clutch hitters exist? *Baseball Research Journal 2.*

Croskey, M.A., P.M. Dawson, A.C. Luessen, I.E. Marohn, and H.E. Wright (1922). The height of the center of gravity of man. *American Journal of Physiology 61*, 171–185.

Crowder, M., M. Dixon, A. Ledford, and M. Robinson (2002). Dynamic modelling and prediction of English Football League matches for betting. *The Statistician 51*, 157–168.

Czarnitzki, D. and G. Stadtmann (2002). Uncertainty of outcome versus reputation: Empirical evidence for the first German football division. *Empirical Economics 27*, 101–112.

Dart, J. and J. Gross (2006, April 19). Does it really matter where you finish in the play-offs? *The Guardian.*

Dawson, P., S. Dobson, J. Goddard, and J. Wilson (2007). Are football referees really biased and inconsistent? Evidence on the incidence of disciplinary sanctions in the English Premier League. *Journal of the Royal Statistical Society Series A 170*, 231–250.

De Boer, R.W., J. Cari, W. Vaes, J.P. Clarijs, A.P. Hollander, G. De Groot, and G.J. Van Ingen Schenau (1987). Moments of force, power, and muscle coordination in speed skating. *International Journal of Sports Medicine 8*(6), 371–378.

De Boer, R.W., G.J. Ettema, H. Van Gorkum, G. De Groot, and G.J. Van Ingen Schenau (1987). Biomechanical aspects of push off techniques in speed skating the curves. *International Journal of Sports Biomechanics 3*, 69–79.

De Koning, J.J., G. De Groot, and G.J. Van Ingen Schenau (1989). Mechanical aspects of the sprint start in olympic speed skating. *International Journal of Sports Biomechanics 5*, 151–168.

De Koning, J.J., G. De Groot, and G.J. Van Ingen Schenau (1991). Coordination of leg muscles during speed skating. *Journal of Biomechanics 24*(2), 137–146.

Deakin, M.A.B. (1967). Estimating bounds on athletic performance. *Mathematics Gazette 51*, 100–103.

Dixon, M.J. and S.C. Coles (1997). Modelling association football scores and inefficiencies in the football betting market. *Applied Statistics 46*, 265–280.

Dixon, M.J. and P.F. Pope (2004). The value of statistical forecasts in the UK association football betting market. *International Journal of Forecasting 20*, 697–711.

Dixon, M.J. and M.E. Robinson (1998). A birth process for association football matches. *The Statistician 47*, 523–538.

Dobson, S. and J. Goddard (2001). *The Economics of Football*. Cambridge: Cambridge University Press.

Duan, N. (1983). Smearing estimate: A nonparametric retransformation method. *Journal of the American Statistical Association 78*, 605–610.

Dubin, C.L. (1990). *Commission of Inquiry into the Use of Drugs and Banned Practices Intended to Increase Athletic Performance*. Ottawa: Canadian Government Publishing Center.

Dyte, D. and S.R. Clarke (2000). A ratings based Poisson model for World Cup simulation. *Journal of the Operational Research Society 51*, 993–998.

Efron, B. and C. Morris (1977). Stein's paradox in statistics. *Scientific American 236*(5), 119–127.

Fizel, J. (2005). The National Football League conundrum. In J. Fizel (Ed.), *The Handbook of Sports Economics Research*, pp. 170–171. M.E. Sharpe, Inc.

Forrest, D., J. Beaumont, J. Goddard, and R. Simmons (2005). Home advantage and the debate about competitive balance in professional sports leagues. *Journal of Sports Sciences 23*(4), 439–445.

Forrest, D., J. Goddard, and R. Simmons (2005). Odds setters as forecasters. *International Journal of Forecasting 21*, 551–564.

Forrest, D. and R. Simmons (2000a). Forecasting sport: the behaviour and performance of football tipsters. *International Journal of Forecasting 16*, 316–331.

Forrest, D. and R. Simmons (2000b). Making up the results: the work of the Football Pools Panel, 1963-1997. *The Statistican 49*(2), 253–260.

Forrest, D. and R. Simmons (2002). Outcome uncertainty and attendance demand in sport: The case of English soccer. *The Statistician 51*(1), 13–38.

Forrest, D., R. Simmons, and B. Buraimo (2005). Outcome uncertainty and the couch potato audience. *Scottish Journal of Political Economy 52*, 641–661.

Fox, D. (2005, November 10). Tony LaRussa and the search for significance. http://www.hardballtimes.com/main/article/tony-larussa-and-the-search-for-significance.

Francis, A.W. (1943). Running records. *Science 98*, 315–316.

Gale, D. (1980). Optimal strategy for serving in tennis. *Mathematics Magazine 44*, 197–199.

Garcia, J. and P. Rodriguez (2002). The determinants of football match attendance revisited: Empirical evidence from the Spanish football league. *Journal of Sports Economics 3*(1), 18–38.

Gardner, P. (2005). DC United outdo Adu. *World Soccer*, 22–23.

Gelman, A., J.B. Carlin, H.S. Stern, and D.B. Rubin (2003). *Bayesian Data Analysis* (2nd ed.). Boca Raton: CRC Press/Chapman & Hall.

George, S.L. (1973). Optimal strategy in tennis: A simple probabilistic model. *Applied Statistics 22*, 97–104.

Gillman, L. (1985). Missing more serves may win more points. *Mathematics Magazine 58*, 222–224.

Goddard, J. (2005). Regression models for forecasting goals and match results in association football. *International Journal of Forecasting 21*, 331–340.

Goddard, J. and I. Asimakopoulos (2004). Forecasting football match results and the efficiency of fixed-odds betting. *Journal of Forecasting 23*, 51–66.

Goddard, J. and S. Thomas (2006). The efficiency of the UK fixed-odds betting market for Euro 2004. *International Journal of Sports Finance 1*, 21–32.

Gompertz, B. (1825). On the nature of the function expressive of the law of human mortality, and on a new mode of determining life contingencies. *Philosophical Transactions of the Royal Society of London 115*, 513–585.

Grabiner (1993). Clutch hitting study. http://www.baseball1.com/bb-data/grabiner/fullclutch.html.

Greene, W.H. (2003). *Econometric Analysis* (5th ed.). Prentice Hall.

Griffiths, R.C. and R.K. Milne (1978). A class of bivariate Poisson processes. *Journal of Multivariate Analysis 8*, 380–395.

Grubb, H.J. (1998). Models for comparing athletic performances. *The Statistician 47*, 509–521.

Haan, M.A., R.H. Koning, and A. van Witteloostuijn (2007). The effects of institutional change in European soccer. Mimeo, University of Groningen, The Netherlands.

Hannan, E.L. (1976). An analysis of different serving strategies in tennis. In R.E. Machol, S.P. Ladany, and D.G. Morrison (Eds.), *Management Science in Sports*, pp. 125–135. New York: North-Holland.

Hart, R., J. Hutton, and T. Sharot (1975). A statistical analysis of association football attendance. *Journal of the Royal Statistical Society, Series C 24*, 17–27.

Hastie, T., R. Tibshirani, and J. Friedman (2001). *The Elements of Statistical Learning*. New York: Springer.

Hellebrandt, F.A. and E.B. Franssen (1943). Physiological study of the vertical standing of man. *Physiology Review 23*, 220–255.

Hill, A.V. (1913). The combinations of haemoglobin with oxygen and with carbon monoxide. *Biochemistry 7*, 471–480.

Hodak, G.A. (1988). Gordon H. Adam, 1936 Olympic Games. Olympic Oral History Project, Amateur Athletic Foundation of Los Angeles.

Howard, R.A. (1960). *Dynamic Programming and Markov Processes*. Cambridge, MA: MIT Press.

Insley, R., L. Mok, and T.B. Swartz (2004). Issues related to sports gambling. *The Australian and New Zealand Journal of Statistics 46*, 219–232.

Iwaoka, K., H. Hatta, Y. Atomi, and M. Miyashita (1988). Lactate, respiratory compensation thresholds, and distance running performance in runners of both sexes. *International Journal of Sports Medicine 9*(5), 306–309.

Jackson, D.A. (1989). Letter to the editor on "Probability models for tennis scoring systems" by L.H. Riddle. *Applied Statistics 38*, 377–378.

Jacoby, E. and B. Fraley (1995). *Complete Book of Jumps*. Champaign, IL: Human Kinetics.

James, B. (1984). *The Bill James Baseball Abstract*. New York: Ballantine Books.

James, B. (2006). *The Bill James Handbook 2007*. Skokie, IL.: ACTA Sports.

Jang, K.T., M.G. Flynn, D.L. Costill, J.P. Kirwin, J.A. Houmard, J.B. Mitchell, and L.J. D'Acquisto (1987). Energy balance in competitive swimmers and runners. *Journal of Swimming Research 3*, 19–23.

Janoschek, A. (1957). Das reaktionskinetische Grundgesetz und seine Beziehungen zum Wachstums- und Ertragsgesetz. *Statistische Vierteljahresschrift 10*, 25–37.

Karlis, D. and I. Ntzoufras (2003). Analysis of sports data by using bivariate Poisson models. *The Statistician 52*, 381–393.

Kennedy, P. (2004). *A Guide to Econometrics* (5th ed.). Oxford (UK): Blackwell Publishing.

Kennelly, A.E. (1905). *A study of racing animals*. The American Academy of Arts and Sciences.

Kingston, J.G. (1976). Comparison of scoring systems in two-sided competitions. *Journal of Combinatorial Theory 20*, 357–362.

Klaassen, F.J.G.M. and J.R. Magnus (2001). Are points in tennis independent and identically distributed? Evidence from a dynamic binary panel data model. *Journal of the American Statistical Association 96*, 500–509.

Klaassen, F.J.G.M. and J.R. Magnus (2006). Are economic agents successful maximizers? An analysis through service strategy in tennis. Discussion Paper 2006-52, CentER, University of Tilburg, Tilburg, The Netherlands. Submitted for publication.

Koning, R.H. (2000). Balance in competition in Dutch soccer. *The Statistician 49*(3), 419–431.

Koop, G. (2004). Modelling the evolution of distributions: An application to Major League Baseball. *Journal of the Royal Statistical Society, Series A 167*, 639–655.

Kuper, G.H. and E. Sterken (2003). Endurance in speed skating: The development of world records. *European Journal of Operational Research 148*(2), 293–301.

Kuypers, T. (1996). The beautiful game? An econometric study of why people watch English football. Discussion Paper 96-01, Department of Economics, University College London, London.

Kuypers, T. (1997). *The beautiful game? An econometric study of audiences, gambling, and efficiency in English football.* Ph. D. thesis, University College London, London.

Kuypers, T. (2000). Information and efficiency: an empirical study of a fixed odds betting market. *Applied Economics 32*, 1353–1363.

Lane, F.C. (1925). *Batting.* Cleveland, Ohio: Society for American Baseball Research.

Lee, A. (1999). Modelling rugby league data via bivariate negative binomial regression. *Australian and New Zealand Journal of Statistics 41*, 153–171.

Lee, Y.H. and D.J. Berri (in press). A re-examination of production functions and efficiency estimates for the National Basketball Association. *Scottish Journal of Political Economy.*

Lehmann, E.L. and G. Casella (2003). *Theory of Point Estimation* (2nd ed.). New York: Springer.

Lerner, L. (1996). *Physics for scientists and engineers.* Boston: Jones and Bartlett.

Lewis, M. (2004). *Moneyball: The Art of Winning an Unfair Game.* New York: W.W. Norton & Company.

Lietzke, M.H. (1954). An analytical study of world and Olympic racing records. *Science 119*, 333–336.

Linhart, H. and W. Zucchini (1986). *Model Selection.* New York: John Wiley and Sons.

Linthorne, N.P. (1999, 31 October - 5 November). Optimum throwing and jumping angles in athletics. In *Proceedings 5th IOC Congress on Sport Sciences with Annual Conference of Science and Medicine in Sport*, Sydney, Australia.

Little, A. (1995). *Wimbledon Compendium 1995.* London: All England Lawn Tennis and Croquet Club.

Loland, R. (2001). *Fair Play in Sport.* London: Routledge.

Long, J.S. and J. Freese (2003). *Regression Models for Categorical Dependent Variables using Stata.* College Station, TX: Stata Press.

Lutoslawska, G., A. Klusiewics, D. Sitkowsi, and B. Krawczyk (1996). The effect of simulated 2 km laboratory rowing on blood lactate, plasma inorganic phosphate, and ammonia in male and female junior rowers. *Biology of Sport 13*(1), 31–38.

Mabel's Tables (2003). Football yearbook. `http://www.mabels-tables.com`.

MacKinnon, J., H. White, and R. Davidson (1983). Tests for model specification in the presence of alternative hypothesis: Some further results. *Journal of Econometrics 21*, 53–70.

Maddala, G.S. (2001). *Introduction to Econometrics* (3rd ed.). New York: John Wiley.

Magnus, J.R. and F.J.G.M. Klaassen (1999a). The effect of new balls in tennis: Four years at Wimbledon. *The Statistician 48*, 239–246.

Magnus, J.R. and F.J.G.M. Klaassen (1999b). The final set in a tennis match: Four years at Wimbledon. *Journal of Applied Statistics 26*, 461–468.

Magnus, J.R. and F.J.G.M. Klaassen (1999c). On the advantage of serving first in a tennis set: Four years at Wimbledon. *The Statistician 48*, 247–256.

Maher, M.J. (1982). Modelling association football scores. *Statistica Neerlandica 36*, 109–118.

Maisel, H. (1966). Best k of $2k - 1$ comparison. *Journal of the American Statistical Association 61*, 329–344.

Malthus, T.R. (1798). *An Essay on the Principle of Population.* London: J. Johnson. Available at `http://www.econlib.org/Library/Malthus/malPop.html`.

Manning, W.G. and J. Mullahy (2001). Estimating log models: To transform or not to transform? *Journal of Health Economics 20*, 461–494.

Maud, P.J. and B.B. Schultz (1986). Gender comparisons in anaerobic power and anaerobic capacity tests. *British Journal of Sports Medicine 20*(2), 51–54.

McArdle, W., F.I. Katch, and V.L. Katch (1981). *Exercise Physiology*. Philadelphia: Lea and Febiger.

McCracken, V. (2001). Pitching and defense: how much control do hurlers have? Baseball Prospectus. http://www.baseballprospectus.com.

McCullagh, P. and J.A. Nelder (1989). *Generalized Linear Models* (2nd ed.). Monographs on statistics and applied probability 37. London: Chapman and Hall.

McFadden, D. (1984). Econometric analysis of qualitative choice models. In Z. Griliches and M.D. Intriligator (Eds.), *Handbook of Econometrics*, Volume II, Chapter 24. Amsterdam: North-Holland.

McGinnis, P.M. (1991). Biomechanics and pole vaulting: Run fast and hold high. In *Proceedings of the American Society of Biomechanics, 15th Annual Meeting*, Tempe, AZ, pp. 16–17.

McGinnis, P.M. (1997). Mechanics of the pole vault take-off. *New Studies in Athletics 12*(1), 43–46.

McHale, I.G. and P.A. Scarf (2006). Modelling soccer matches using bivariate discrete outcomes. Technical report, University of Salford, Salford, United Kingdom.

McQuarrie, A.D.R. and C.-L. Tsai (1998). *Regression and Time Series Model Selection*. Singapore: World Scientific.

Miles, R.E. (1984). Symmetric sequential analysis: The efficiencies of sports scoring systems (with particular reference to those of tennis). *Journal of the Royal Statistical Society, Series B 46*, 93–108.

Morrison, D.G. and M.U. Kalwani (1993). The best NFL field goal kickers: Are they lucky or good? *Chance 6*(3), 30–37.

Mosteller, F. and J.W. Tukey (1977). *Data Analysis and Regression*. Reading, MA.: Addison-Wesley.

Nevill, A.M., N.J. Balmer, and A.M. Williams (2002). The influence of crowd noise and experience upon refereeing decisions in football. *Psychology of Sport and Exercise 3*, 261–272.

Nevill, A.M and G. Whyte (2005). Are there limits to running world records? *Medicine and Science in Sports and Exercise 37*, 1785–1788.

NFL Record & Fact Book (Various editions). http://www.nfl.com/history/randf.

Noll, R.G. (2003). The organization of sports leagues. *Oxford Review of Economic Policy 19*(4), 530–551.

Norman, J.M. (1985). Dynamic programming in tennis — when to use a fast serve. *Journal of Operational Research Society 36*, 75–77.

Paton, D. and A. Cooke (2005). Attendance at county cricket: An economic analysis. *Journal of Sports Economics 6*, 24–45.

Peel, D.A. and D.A. Thomas (1988). Outcome uncertainty and the demand for football. *Scottish Journal of Political Economy 35*, 242–249.

Peel, D.A. and D.A. Thomas (1992). The demand for football: Some evidence on outcome uncertainty. *Empirical Economics 17*, 323–331.

Peel, D.A. and D.A. Thomas (1996). Attendance demand: An investigation of repeat fixtures. *Applied Economics Letters 3*(6), 391–394.

Pollard, R. (1986). Home advantage in soccer: A retrospective analysis. *Journal of Sports Sciences 4*, 237–248.

Pollard, R. (2002). Evidence of a reduced home advantage when a team moves to a new stadium. *Journal of Sports Sciences 20*(12), 969–973.

Pollard, R. (2006). Worldwide regional variations in home advantage in association football. *Journal of Sports Sciences 24*(3), 231–240.

Pope, P.F. and D.A. Peel (1989). Information, prices, and efficiency in a fixed-odds betting market. *Economica 56*, 323–341.

Quirk, J. and R.D. Fort (1992). *Pay Dirt: The Business of Professional Team Sports*. Princeton, NJ: Princeton University Press.

Rabe-Hesketh, S., A. Skrondal, and A. Pickles (2004). *GLLAMM Manual*. U.C. Berkeley Division of Biostatistics Working Paper Series.

Rascher, D. (1999). A test of the optimal positive production network externality in Major League Baseball. In J. Fizel, E. Gustafson, and L. Hadley (Eds.), *Sports Economics: Current Research*, Westport, CT. Praeger.

Ratkowsky, D.A. (1990). *Handbook of Nonlinear Regression Models*. New York: Marcel Dekker.

Reep, R. and B. Benjamin (1968). Skill and chance in Association Football. *Journal of the Royal Statistical Society A 131*, 581–585.

Richards, F.J. (1959). A flexible growth function for empirical use. *Journal of Experimental Botany 10*, 290–300.

Ridder, G., J.S. Cramer, and P. Hopstaken (1994). Down to ten: Estimating the effect of a red card in soccer. *Journal of the American Statistical Association 89*, 1124–1127.

Riddle, L.H. (1988). Probability models for tennis scoring systems. *Applied Statistics 37*, 63–75 and 490.

Riddle, L.H. (1989). Author's reply to D.A. Jackson. *Applied Statistics 38*, 378–379.

Ruane, T. (2005). In search of clutch hitting. `http://retrosheet.org/Research/RuaneT/clutch_art.htm`.

Rue, H. and O. Salvesen (2000). Prediction and retrospective analysis of soccer matches in a league. *The Statistician 49*, 399–418.

Schabenberger, O., B.E. Tharp, J.J. Kells, and D. Penner (1999). Statistical tests for Hormesis and effective dosages in herbicide dose response. *Agronomy Journal 91*, 713–721.

Schnute, J. (1981). A versatile growth model with statistically stable parameters. *Canadian Journal of Fishery and Aquatic Science 38*, 1128–1140.

Scully, G.W. (1974). Pay and performance in major league baseball. *American Economic Review 64*, 915–930.

Silver, N. (2004). Lies, damned lies: The unique Ichiro. Baseball Prospectus. `http://www.baseballprospectus.com`.

Silver, N. (2006). Is David Ortiz a clutch hitter? In J. Keri (Ed.), *Baseball Between the Numbers*, Chapter 1-2, pp. 14–34. New York: Basic Books.

Simon, G.A. and J.S. Simonoff (2006). "Last licks": Do they really help? *The American Statistician 60*(1), 13–18.

Simonoff, J.S. (2003). *Analyzing Categorical Data*. New York: Springer-Verlag.

Skrondal, A. and S. Rabe-Hesketh (2004). *Generalized Latent Variable Modeling: Multilevel, Longitudinal, and Structural Equation Models*. Boca Raton: Chapman & Hall/CRC.

Smith, R.L. (1988). Forecasting records by maxiumum likelihood. *Journal of the American Statistical Association 83*, 331–388.

StataCorp (2003). *Stata Statistical Software: Release 8*. College Station, TX: Stata-Corp LP.

Statmail (2003). Statmail. `http://www.statmail.co.uk`.

Stefani, R.T. (1987). Applications of statistical methods to American Football. *Journal of Applied Statistics 14*(1), 61–73.

Stefani, R.T. (1998). Predicting outcomes. In J. Bennett (Ed.), *Statistics in Sport*, Chapter 12. New York: Oxford University Press.

Stefani, R.T. (1999). A taxonomy of sports rating systems. *IEEE Trans. On Systems, Man and Cybernetics, Part A 29*(1), 116–120.

Stefani, R.T. (2006). The relative power output and relative lean body mass of world and Olympic male and female champions with implications for gender equity. *Journal of Sports Sciences 24*(12), 1329–1339.

Stefani, R.T and S.R. Clarke (1992). Predictions and home advantage for Australian rules football. *Journal of Applied Statistics 19*(2), 251–259.

Stefani, R.T. and D. Stefani (2000). Power output of Olympic rowing champions. *Olympic Review XXVI-30*, 59–63.

Sterken, E. (2005). A stochastic frontier approach to running performance. *IMA Journal of Management Mathematics 16*, 141–149.

Stern, H.S. and C.N. Morris (1993). Looking for small effects: power and finite sample bias considerations. (Comment on C. Albright's "A statistical analysis of hitting streaks in baseball"). *Journal of the American Statistical Association 88*, 1189–1194.

Summers, A. (2005). Hockey pools for profit: a simulation based player selection strategy. Master's thesis, Department of Statistics and Actuarial Science, Simon Fraser University, Burnaby BC, Canada.

Szymanski, S. (2003). The economic design of sporting contests. *Journal of Economic Literature 41*, 1137–1187.

Szymanski, S. and A. Zimbalist (2003). *National Pastime*. Washington DC: Brookings Institution Press.

Tatem, A.J., C.A. Guerra, P.M. Atkinson, and S.I. Hay (2004). Momentous sprint at the 2156 Olympics? *Nature 431*, 525.

Thorn, J. and P. Palmer (1984). *The Hidden Game of Baseball: A Revolutionary Approach to Baseball and Its Statistics*. New York: Doubleday.

Torgler, B. (2004). The economics of the FIFA football World Cup. *Kyklos 57*, 287–300.

Toussaint, H.M. and P.J. Beck (1992). Biomechanics of competitive front crawl swimming. *Sports Medicine 13*, 8–24.

Toussaint, H.M., G. De Groot, H.H. Savelberg, K. Vervoorn, A.P. Hollander, and G.J. Van Ingen Schenau (1988). Active drag related to velocity in male and female swimmers. *Journal of Biomechanics 21*, 435–438.

Toussaint, H.M., T.Y. Janssen, and M. Kluft (1991). Effect of propelling surface size on the mechanics and energetics of front crawl swimming. *Journal of Biomechanics 24*, 205–211.

Toussaint, H.M., W. Knops, G. De Groot, and A.P. Hollander (1990). The mechanical efficiency of front crawl swimming. *Medicine and Science in Sports and Exercise 22*, 402–408.

Tuck, L.O. and L. Lazauskas (1996, 30 Sept.- 2 Oct.). Low drag rowing shells. In *Third Conference on Mathematics and Computers in Sport*, Queensland, Australia, pp. 17–34. Bond University.

Van Ingen Schenau, G.J., G. De Groot, and R.W. De Boer (1985). The control of speed in elite female speed skaters. *Journal of Biomechanics 18*(2), 91–96.

Vrooman, J. (1996). The baseball's player market reconsidered. *Southern Economic Journal 63*(2), 339–360.

Wainer, H., C. Njue, and S. Palmer (2000). Assessing time trends in sex differences in swimming and running. *Chance 13(1)*, 10–21.

Wakai, M. and N.P. Linthorne (2000, 7-12 September). Optimum takeoff angle in the standing long jump. In *Sports Medicine Book of Abstracts, 2000 Pre-Olympic Congress: International Congress on Sport Science, Sports Medicine, and Physical Education*, Brisbane, Australia.

Wallechinsky, D. (2004). *The Complete Book of the Summer Olympics, Athens 2004 Edition*. Sport Classic.

Wallechinsky, D. (2005). *The Complete Book of the Winter Olympics, Turin 2006 Edition*. Sport Classic.

Welki, A. and T. Zlatoper (1999). U.S. professional football game-day attendance. *Atlantic Economic Journal 27*, 285–298.

Wellock, I.J., G.C. Emmans, and I. Kyriazakis (2004). Describing and predicting potential growth in the pig. *Animal Science 78*, 379–388.

Weyand, P.G., D.B. Sternlight, M.J. Bellizzi, and S. Wright (2000). Faster top running speeds are achieved with greater ground forces not rapid leg movements. *Journal of Applied Physiology 89*, 1991–1999.

Whipp, B.J. and S.A. Ward (1992). Will women soon outrun men? *Nature 355*, 25.

Woolner, K. (2001). Counterpoint: Pitching and defense. Baseball Prospectus. http://www.baseballprospectus.com.

List of authors

Contact addresses for the authors are

Jim Albert, Department of Mathematics and Statistics, Bowling Green State University, Bowling Green, OH 43403, USA,
email: albert@bgnet.bgsu.edu.

David J. Berri, Department of Applied Economics, California State University-Bakersfield, USA,
email: dberri@csub.edu.

Chris Bojke, Pharmerit UK, Suite 8, Tower House, Fishergate, York YO10 4UA, United Kingdom,
email: cbojke@pharmerit.com.

Babatunde Buraimo, Lancashire Business School, University of Central Lancashire, Preston PR1 2HE, United Kingdom,
email: baburaimo@uclan.ac.uk.

Stephen Davies Salford, Business School, The University of Salford, Salford M5 4WT, United Kingdom, email: stephenmdavies1983@yahoo.co.uk.

Stephen Dobson, Department of Economics, University of Otago, PO Box 56, Dunedin, New Zealand,
email: sdobson@business.otago.ac.nz.

David Forrest, Centre for the Study of Gambling, University of Salford, Salford M5 4WT, United Kingdom,
email: D.K.Forrest@salford.ac.uk.

John Goddard, Bangor Business School, University of Wales, Bangor, Gwynedd, LL57 2DG, United Kingdom,
email: j.goddard@bangor.ac.uk.

Marco Haan, Institute of Economics and Econometrics, University of Groningen, PO Box 800, 9700 AV Groningen, The Netherlands,
email: m.a.haan@rug.nl.

Franc J.G.M. Klaassen, Department of Economics, University of Amsterdam, Roetersstraat 11, 1018 WB Amsterdam, The Netherlands,
email: f.klaassen@uva.nl.

Ruud H. Koning, Institute of Economics and Econometrics, University of Groningen, PO Box 800, 9700 AV Groningen, The Netherlands,
email: r.h.koning@rug.nl.

Gerard H. Kuper, Institute of Economics and Econometrics, University of Groningen, PO Box 800, 9700 AV Groningen The Netherlands,
email: g.h.kuper@rug.nl.

Richard Lockhart, Department of Statistics and Actuarial Science, Simon Fraser University, 8888 University Drive, Burnaby BC, Canada V5A156,
email: lockhart@stat.sfu.ca.

Jan R. Magnus, Department of Econometrics & Operations Research, Tilburg University, P.O. Box 90153, 5000 LE Tilburg, The Netherlands,
email: magnus@uvt.nl.

Ian G. McHale, Centre for Operational Research and Applied Statistics, The University of Salford, Salford M5 4WT, United Kingdom,
email: I.McHale@salford.ac.uk.

Rebecca J. Sela, Leonard N. Stern School of Business, New York University, 44 West 4th Street, New York, NY 10012 USA,
email: rsela@stern.nyu.edu.

Jeffrey S. Simonoff, Leonard N. Stern School of Business, New York University, 44 West 4th Street, New York, NY 10012 USA,
email: jsimonof@stern.nyu.edu.

Robert Simmons, Department of Economics, The Management School, Lancaster University, Lancaster LA1 4YX, United Kingdom,
email: r.simmons@lancaster.ac.uk.

Raymond T. Stefani, 25032 Via Del Rio, Lake Forest, CA 92630 USA,
email: stefani@csulb.edu.

Elmer Sterken, Institute of Economics and Econometrics, University of Groningen, PO Box 800, 9700 AV Groningen, The Netherlands,
email: e.sterken@rug.nl.

Hal S. Stern, Department of Statistics, University of California, Irvine, CA 92697-1250, USA,
email: sternh@uci.edu.

Adam Sugano, Department of Statistics, University of California, Los Angeles, CA 90024, USA,
email: asugano@ucla.edu.

Amy Summers, Syreon Corporation, 450-1385 West 8th Avenue, Vancouver BC, Canada V6H3V9,
email: asummer1@sfu.ca.

Tim Swartz, Department of Statistics and Actuarial Science, Simon Fraser University, 8888 University Drive, Burnaby BC, Canada V5A156,
email: `tim@stat.sfu.ca`.

Arjen van Witteloostuijn, Department of International Economics and Business, University of Groningen, PO Box 800, 9700 AV Groningen, The Netherlands, email: `a.van.witteloostuijn@rug.nl`.

Index